It is widely recognised that men and women in societies all over the world have very different experiences of sickness and health. This collection brings together biological and social anthropologists whose work illustrates how these sub-disciplines have approached the task of explaining such differences. We demonstrate that an understanding of science and culture, using the notions of biological 'sex' and socio-culturally constructed 'gender' are both essential both furthering analyses of men's and women's, boys' and girls' experiences of health and disease. We address the important topics of gender differences in parental care, cardiovascular disease, reproductive health and psychological illness, and look at how the medicalisation of women and their relative absence from models of population health might affect their experiences of preventative health measures. This book will be particularly useful for students on human sciences or anthropology courses, or anyone wishing to gain an interdisciplinary perspective on the subject.

TESSA M. POLLARD is a Lecturer in Anthropology at the University of Durham, UK. She is interested in the ways in which biological and socially constructed differences between men and women cause their risks of disease to differ, especially with respect to the effects of psychosocial stress on health.

SUSAN BRIN HYATT is Assistant Professor of Anthropology at Temple University, Philadelphia. She has written on women's engagement in local-level activism on council estates in England, and is also interested in comparative perspectives on welfare policies and poverty in the US and the UK.

THE BIOSOCIAL SOCIETY SYMPOSIUM SERIES

Series editor: Professor G. A. Harrison, University of Oxford

The aim of the Biosocial Society is to examine topics and issues of biological and social importance and to promote studies of biosocial matters. By examining various contemporary issues and phenomena, which clearly have dimensions in both the social and biological sciences, the society hopes to foster the integration and inter-relationships of these dimensions.

Previously published volumes

1. Famine *edited by G. A. Harrison*
2. Biosocial Aspects of Social Class *edited by C. G. N. Mascie-Taylor*
3. Mating and Marriage *edited by V. Reynolds and J. Kellet*
4. Social & Biological Aspects of Ethnicity *edited by M. Chapman*
5. The Anthropology of Disease *edited by C. G. N. Mascie-Taylor*
6. Human Adaptation *edited by G. A. Harrison*
7. Health Interactions in Less-developed Countries *edited by S. J. Ulijaszek*
8. Health Outcomes: Biological, Social & Economic Perspectives *edited by H. Macbeth*
9. The Anthropology of War *edited by M. Parker*
10. Biosocial Perspectives on Children *edited by C. Panter-Brick*

Volumes 1–9 are available from Oxford University Press

Sex, Gender and Health

Edited by

TESSA M. POLLARD
University of Durham, UK

and

SUSAN BRIN HYATT
Temple University, Philadelphia

PUBLISHED BY THE PRESS SYNDICATE OF THE UNIVERSITY OF CAMBRIDGE
The Pitt Building, Trumpington Street, Cambridge, United Kingdom

CAMBRIDGE UNIVERSITY PRESS
The Edinburgh Building, Cambridge CB2 2RU, UK www.cup.cam.ac.uk
40 West 20th Street, New York, NY 10011–4211, USA www.cup.org
10 Stamford Road, Oakleigh, Melbourne 3166, Australia
Ruiz de Alarcón 13, 28014 Madrid, Spain

First published 1999

Printed in the United Kingdom at the University Press, Cambridge

Typeset in Monotype Baskerville 11½/14pt, in QuarkXPress™ [WV]

A catalogue record for this book is available from the British Library

Library of Congress Cataloguing in Publication data

Sex, gender, and health/edited by Tessa M. Pollard and Susan B. Hyatt.
p. cm.
Includes bibliographical references and index.
ISBN 0 521 59282 8. – ISBN 0 521 59707 2 (pbk.)
1. Health – Sex differences. 2. Medical anthropology. 3. Health – Cross-cultural studies. I. Pollard, Tessa M., 1966– . II. Hyatt, Susan B., 1953– .
RA564.7.S49 1999
613–dc21 98-32165 CIP

ISBN 0 521 59282 8 hardback
ISBN 0 521 59707 2 paperback

Contents

Contributors

HELEN L. BALL
Department of Anthropology, University of Durham,
43 Old Elvet, Durham, DH1 3HN, UK

CATHERINE M. HILL
Department of Anthropology, University of Durham,
43 Old Elvet, Durham, DH1 3HN, UK

SUSAN BRIN HYATT
Department of Anthropology, Temple University,
Gladfelter Hall, Philadelphia, PA 19122, USA

PATRICIA A. KAUFERT
Department of Community Health Sciences,
University of Manitoba, 750 Bannantyne Avenue, Winnipeg,
Manitoba, R3E OW3, Canada

ROLAND LITTLEWOOD
Department of Anthropology, University College London,
Gower Street, London WC1E 6BT, UK

LENORE MANDERSON
Key Centre for Women's Health, University of Melbourne,
720 Swanston Street, Carlton, Vic 3053, Australia

TESSA M. POLLARD
Department of Anthropology, University of Durham,
43 Old Elvet, Durham, DH1 3HN, UK

EMILY K. ROUSHAM
Department of Human Sciences, Loughborough University,
Loughborough, Leicestershire LE11 3TU, UK

Preface

This book is the eleventh in the series produced by the Biosocial Society, all of which aim to address topics which benefit from the contribution of both biological and social scientists, and particularly from communication between them. It is a product of the Society's 1997 workshop which was held at the University of Durham's Stockton Campus, but also includes some additional contributors. We would like to thank the Society for its support of the meeting and of this volume.

We hope that the book demonstrates that an understanding of science and society, in this case sex and gender, are both essential for furthering our analyses of men's and women's experiences of health and disease, and that reading it is as informative and enlightening an experience as editing it has been.

TMP and SBH
Durham
1998

1

Sex, gender and health: integrating biological and social perspectives

TESSA M. POLLARD AND SUSAN BRIN HYATT

From male and female to men and women

The primary aim of this volume is to show how both biological and social anthropologists have sought to explain the reasons that underlie well-documented differences in the health experiences of men and women. By improving our understanding of the origins of the differences in health experiences between men and women, we hope to achieve greater insight into the processes generating ill-health for everyone. We can also begin to address imbalances in the diagnosis of disease and subsequent treatments which have generally favoured men but which have occasionally advantaged women.

A second aim is to help bring social and biological anthropologists together. Too often they fail to communicate with one another. In particular, we wish to illustrate the extent to which the field of health research is one which benefits immensely from improved co-operation between social and biological scientists. The first step in such a process must be for each sub-discipline to look beyond its own paradigm to acknowledge the value of others. We hope that we have furthered these goals by inviting a diverse group of contributors to examine a selection of important issues regarding the health of men and women. With its emphasis on holism and on examining human cultures from a wide range of analytical perspectives, anthropology is one of the few fields which can claim to address these issues from both biological and social perspectives under a single disciplinary rubric.

Throughout this volume, beginning with its title, we highlight the distinction between 'sex' and 'gender' because these terms refer to two different aspects of the human experience. By 'sex', we mean the specific genetic and hormonal make-up of individuals and their subsequent development of secondary physical characteristics which place individuals in the category 'female' (XX chromosomes) or 'male' (XY chromosomes). At the same time, we also acknowledge that even within the biological category 'sex', there is tremendous variation and there are individuals who, for a variety of reasons, including atypical chromosomal patterns (e.g. XXY), transexualism and intersexuality (hermaphradism), do not fit neatly into either category. With the term 'gender', we refer to a much broader range of variation in how people in societies all over the world understand the social and cultural roles, values and behaviours of men and boys, girls and women. Many social anthropologists use the term 'social construction' to refer to the idea that all aspects of human society and relationships, such as gender, are not 'givens' but are always interpreted through the lens of culture and therefore vary greatly between different times and places.

Most people are born either male or female; men and women, however, are formed as social beings through life-long processes of enculturation. In this book, we are seeking to understand and explain how, in the life course of creating boys and men, girls and women out of males and females, biology and culture interact to produce differences between the sexes in terms of their health experiences and their rates of morbidity and mortality. It is precisely because this interaction between biology and culture is so multifaceted and complex that drawing an absolute distinction between 'sex' and 'gender' in practice is often difficult, if not impossible.

In the majority of countries women now have a longer life expectancy than do men (Figure 1.1), but the difference is particularly marked in affluent countries. Historically this disparity emerged in Europe and North America during the latter part of the nineteenth century and continued to widen during the twentieth century. Now men outlive women only in parts of South Asia.

More detailed examination of the patterns underlying these findings show that boys generally have higher neonatal and infant mortality rates than girls. However, figures from England and

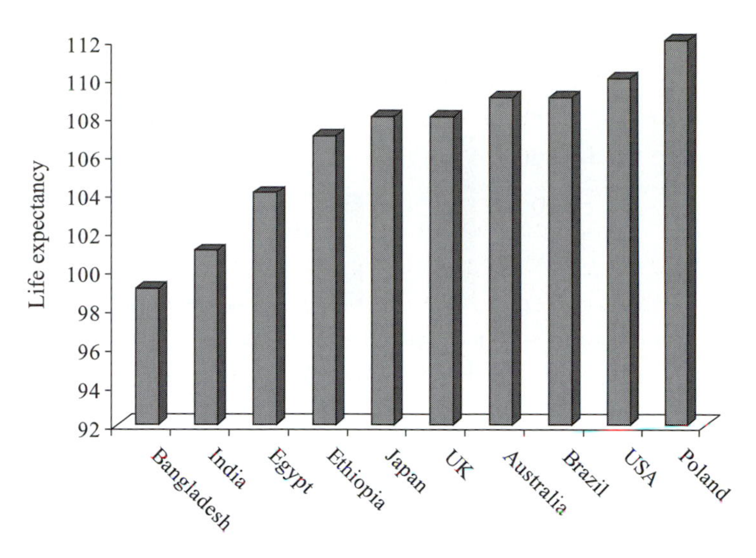

Figure 1.1. Life expectancy of females as a percentage of males in 1990. A figure above 100 indicates that the female average is higher than the male. Data from the United Nations Development Programme (1992).

Wales show that men's mortality disadvantage peaks first in early adulthood and again in men in their sixties (Hart 1989). The first peak results from the greater number of violent and accidental deaths suffered by young men (Figure 1.2). The second is a consequence of the higher mortality of men from the diseases that are responsible for the majority of deaths across industrialised populations – cardiovascular disease and cancers (Figure 1.2). Other European countries show less marked age patterns, but the same general explanations hold. In France, where cardiovascular disease is less predominant than in northern Europe, men show higher mortality than women from cirrhosis of the liver, a disease which is more prevalent than in northern Europe (Hart 1989).

The smaller mortality advantage of women in poorer countries can partly be attributed to deaths associated with childbearing. Around 500 000 women die every year as a result of complications during pregnancy or delivery and many more die from related complications (Holloway 1994). Unsafe abortions also account for a high proportion of these deaths, and in African countries such as

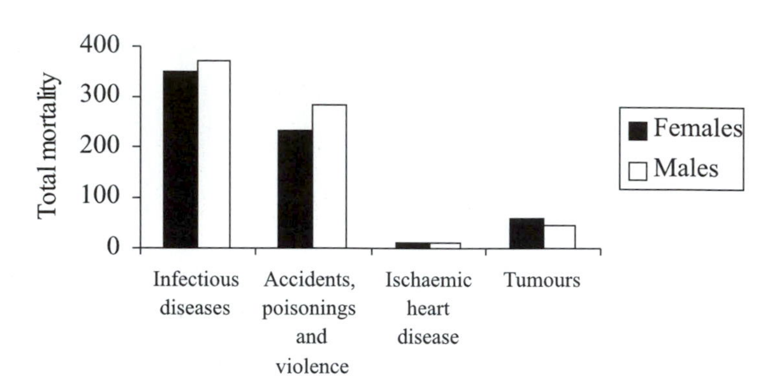
Guatemala 1976
Total mortality
400
300
200
100
0
Infectious diseases
Accidents, poisonings and violence
Ischaemic heart disease
Tumours
Females
Males

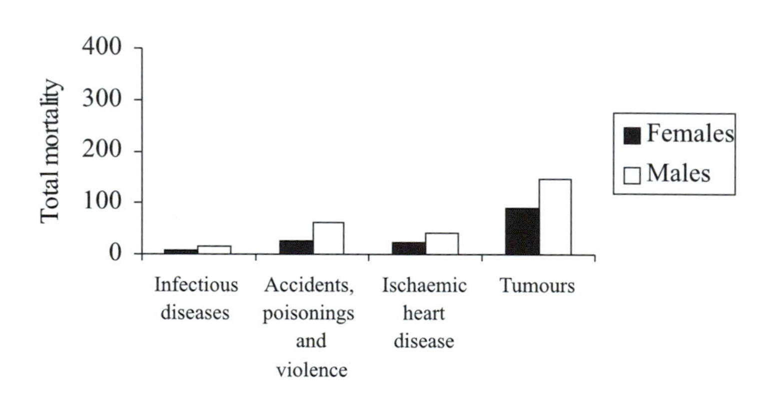
Japan 1978
Total mortality
400
300
200
100
0
Infectious diseases
Accidents, poisonings and violence
Ischaemic heart disease
Tumours
Females
Males

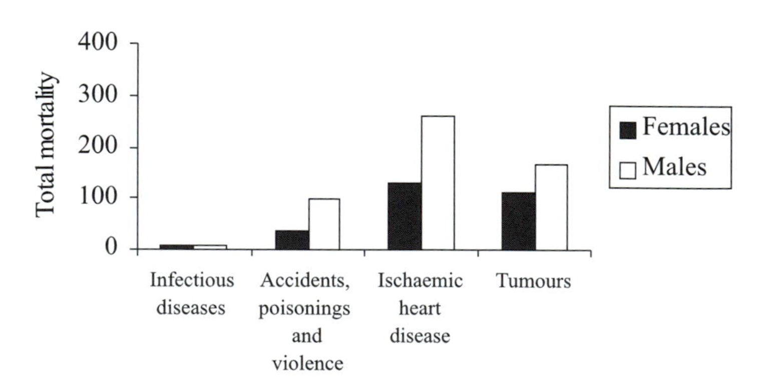
United States 1978
Total mortality
400
300
200
100
0
Infectious diseases
Accidents, poisonings and violence
Ischaemic heart disease
Tumours
Females
Males

Somalia, female circumcision (also known as 'female genital mutilation') contributes to problems during pregnancy and childbirth (Aden *et al.* 1997). Part of the greater improvement in life expectancy generally experienced by women with increasing economic development is certainly attributable to an improvement in maternal mortality rates.

There are many other aspects of women's health and well-being which suffer in poorer countries, but gender differentials in mortality from infectious diseases are not marked in most places. In South Asia, where women have a shorter life expectancy than men, it is generally true that women have a particularly low status (see Rousham, Chapter 3). There is considerable evidence that in these countries boys are favoured over girls even more markedly than in many other cultures. Selective abortion of female foetuses is commonly practised in India and there is evidence that boys are better nourished than girls and that girls are more likely to die as a result of undernutrition and increased susceptibility to infection. It is also true that girls are often less likely to receive medical attention than boys where resources are scarce. In pre-industrial Europe girls also suffered higher mortality than boys, for similar reasons.

In industrialised countries men tend to suffer from more life-threatening illnesses than women, as would be expected given their higher rates of mortality and comparatively shorter life expectancy. However, it is also well-established that women report greater morbidity and appear to suffer more often from physical and mental illnesses, although Macintyre *et al.* (1996) caution against over-generalisations in this regard. In the United States, for example, women experience a higher incidence of acute illnesses and are more likely to have arthritis, chronic sinusitis and digestive problems (Fuller *et al.* 1993). This trend extends to non-western countries such as Thailand (Fuller *et al.* 1993), but less evidence is available for the poorer countries. These findings have sometimes been

Figure 1.2. Total mortality from selected causes for males and females, showing death rates per 100 000 population per year, age-standardised so that comparisons are not affected by differences in age distribution. Data taken from Waldron (1983).

attributed, although usually only partly, to a greater willingness by women to report symptoms, a suggestion that raises important questions of its own.

Seeking explanations

Both biological and social anthropologists have attempted to explain the differences in health outcomes of men and women. Biological anthropologists try to understand biological variation between and within human populations. They undertake their research within the scientific paradigm and often apply an evolutionary understanding to interpret current biological variation.

Waldron (1983) has explored the extent to which genetic differences in the make-up of men and women can explain some of the sex differences in mortality noted above. For example, it is possible that carrying only one X chromosome is a disadvantage to males. The sex hormones may also affect the biological propensity to develop certain diseases. For example, oestrogen is thought to protect against cardiovascular disease and to affect immune functioning. Women show a stronger antibody response to viral illnesses and parasitic infections, but they are also more vulnerable to autoimmune diseases such as rheumatoid arthritis, particularly prior to the onset of menopause (Holden 1987). High levels of testosterone in men have been linked to a greater risk of prostate cancer (Gann *et al.* 1996) and, via the effects of testosterone on serum lipids, of cardiovascular disease.

Biological anthropologists aim to examine how the biology of such processes is affected by variation in the socio-cultural environments of populations and individuals within populations. Their work also extends to the behavioural arena, applying evolutionary theory to understand human behaviour. It is always necessary, then, for biological anthropologists to consider gender, since they aim to understand human biology within the context of human culture. Walker and Cook (1998) have recently drawn attention to this issue, noting that some biological (physical) anthropologists have failed to distinguish between sex and gender in their writing, confusing biological and cultural categories, leading to conceptual confusion and problems in communicating with colleagues in other disciplines. They call on biological anthropologists to use these terms carefully as an important tool in the study of human biology in its cultural context.

Social anthropologists have focused on explanations for variation in health outcomes based on the different gender roles played by boys and girls, men and women in a variety of cultural contexts. As Lorber (1997:2) writes, gender 'creates different risks and protections for physical illnesses, produces different behavior when ill, elicits different responses in health care personnel, affects the social worth of patients, and influences priorities of treatment, research, and financing'. Thus the different roles of men and women may create different health risks. For example, Hart (1989) has suggested that the increasing gap between the survival prospects of men and women seen over the last one hundred years in Europe can be traced to the greater access of the male breadwinner to the economic resources of the household, which allow him to spend disposable income on such luxuries as alcohol and cigarettes. Thus, although men may have had more resources at their disposal, it is partly because of that very fact that they have suffered disproportionately from diseases such as lung cancer and cirrhosis of the liver, which in some cases have been linked to 'lifestyle' issues like smoking and high rates of alcohol consumption. Social processes also contribute to the definition of illness. Thus it has been argued that women have been subjected to greater medicalisation than have men because medical science has based its norms on the exemplary body of the man as the standard against which all others are measured (Urla and Terry 1995, Davis 1996, Lorber 1997). What are normal biological functions for the female (menstruation, menopause, lactation and childbirth, for example) have been treated as essentially medical problems and even as disease-states, even when they proceed without undue complications, rather than being regarded simply as natural functions (see Martin 1987:27–67).

In this volume, we hope to demonstrate that social perspectives and biological science need not find themselves in irreconcilable and antithetical opposition to one another but can, in fact, be used together productively to illustrate the ways in which both contribute to our understanding of the world around us. And, nowhere is it more apparent that each of these perspectives is necessary and useful, we would argue, than in the case of sex and gender. At the most basic level, we believe that it is important for anyone seeking to understand health issues, whether from a biological or a social point of view, to be aware of the insights made possible by both of these approaches. It is for this purpose that we have brought together biological anthropologists who incorporate an understanding of gender into their hypotheses regarding the biological mechanisms that lead to ill-health, as well as social anthropologists who focus on understanding how the interaction between gender and biological factors has affected men and women's physical and mental health outcomes in different societies. The possibilities for dialogue between the two approaches are not always immediately evident, but we argue that awareness is a good first step.

The distinction between 'sex' and 'gender', in fact, rests on the acceptance of the truthfulness of both biological and social perspectives. In Hill and Ball's chapter (Chapter 2) on infanticide in this volume, they demonstrate how some theories to explain the decision of parents to invest differentially in newborn boys and girls have been based on sex, whilst others emphasise the importance of the parents' anticipation of the social roles – that is, gender – to be played by the boy or girl later in life. They show that whilst evolutionary theories based on sex provide fascinating insights and appear to have some limited utility in analysing human populations, it is essential to consider the importance of gender roles when trying to understand sex biases in humans as opposed to other animals. In societies where girls are not seen as potential contributors to the household income, for example, or where impoverished families might be expected to provide large dowries for their daughters when they reach the age of marriage, female infanticide is more likely to occur than in societies where boys and girls are seen as having complementary and equally productive parts to play in sustaining the household economy.

It has been argued that parental investment tends to be biased towards sons throughout childhood in many South Asian populations because of the low status of daughters and women in South Asian societies. Some evidence for such a bias is evident in sex mortality ratios. However, it is important that we develop an understanding of the mechanisms leading to more deaths amongst girls. In Chapter 3 of this volume, Rousham notes that theories about the development of ill-health in girls have focused on greater food allocation to sons at the expense of daughters. In her research she has used anthropometric measures of nutritional status in children taken over a period of months to chart the growth of boys and girls. These measures provide an important insight into the mechanisms that lead to poor health in girls and Rousham is able to link changing economic circumstances to biological outcomes in a manner which provides tangible evidence of the impact of gender. Here, then, scientific methods are used to show how our social understanding of gender is translated directly into different biological outcomes for boys and girls.

Pollard's focus (Chapter 4) on the impact of changing lifestyles on men's and women's cardiovascular health also demonstrates that an understanding of the roles played by both gender and sex is essential for her work. There is a complex interaction between the different biological responses to modernised lifestyles, and both the lifestyle changes and biological responses differ between men and women. For example, research in this field suggests that men's and women's bodies, as biological organisms, respond differently to stressors. The ovarian hormone oestrogen, which is found in much higher levels in women in industrial societies than in more traditional cultures, plays a critical role in protecting women from cardiovascular disease, and its low levels probably play a role in explaining the increased risk of cardiovascular disease in men who have experienced stress. This is a difference at the level of sex. When it comes to people's lived experiences of stress, however, gender also comes into play. Women report more psychological disturbances in their lives as a result of stress than do men. This difference in men's and women's experiences of stress is also a consequence of gender roles; in most industrialised societies, it is women who hold the majority of lower-status jobs and who are most often expected to work outside the home whilst also maintaining their responsibilities as full-time carers within their households. Pollard shows that we need to understand both levels in order to make comprehensive sense of the relationship between lifestyle, including stress, and health.

The remaining chapters focus on the role of gender and the social construction of illness. The rather different analyses presented illustrate the variety of ways in which health differences between men and women may find their origins in social factors. Manderson (Chapter 5), for example, focuses on conditions in developing countries and demonstrates how the lower status of women and their relegation to being valued primarily as bearers of children may translate into a variety of reproductive health problems. Moreover, in addition to her explanation of how reproductive illnesses affect women in developing countries, Manderson also shows how *cultural* attitudes toward reproduction and women's sexuality in a variety of societies may also hinder women from seeking help for reproductive disorders and diseases of the sexual organs. Because their sexuality is often carefully monitored and policed, when illness affects their reproductive organs, women in a variety of settings are often afraid to seek medical attention lest they be accused of violating either local norms of chastity, in the case of unmarried women, or marital fidelity, in the case of married women.

Manderson illustrates how embarrassment and lack of familiarity with the sexual functioning of their own bodies may also inhibit men from seeking help for venereal and other reproductive diseases which affect them. Given the rapid spread of AIDS in the developing world, it is critical that educational programmes which focus on prevention take into account the *social* understandings of sexuality and reproductive health in different cultural contexts. Even as our scientific knowledge of the biological mechanisms through which diseases like AIDS are transmitted continues to grow, Manderson's work makes it all too clear that in order to be truly successful in eradicating the spread of AIDS, and other sexually transmitted diseases, we need to pay *equal* attention to the social dimensions and local cultural understandings of sex, gender and sexuality.

In comparison to Manderson's study of women in developing nations, Hyatt illustrates how social perspectives on poor women in the colonising metropole of England also influenced medical practices. In her chapter (Chapter 6), Hyatt illustrates how social reformers in the nineteenth and twentieth centuries in Britain (and elsewhere) targeted impoverished mothers as the primary causes of negative biological outcomes in their children, such as disease, malnutrition and infant mortality and also of undesirable social consequences, such as juvenile delinquency. Rather than deal with the environmental poverty which compromised the survival of babies and children, many of the measures put into place at that time involved the strict surveillance of poor mothers' conduct. The problems of impoverished mothers became translated into issues of medicine or hygiene to be treated through such innovations as health visiting, whereby mothers were trained in the proper methods of child-rearing. To say that many of these developments led to a view of poor mothers as somehow morally and physically deficient, however, is not to simultaneously deny that many of the changes implemented through campaigns against infant mortality, such as providing clean milk and constructing decent sanitation systems, did not also actually promote much better health. Clearly, such measures did vastly improve the health and welfare of *all* the inhabitants of urban environments. The intense focus on mothers as moral beings however, had a socially constructed outcome as well. One was that 'good mothering' was seen as ultimately perhaps even more essential for fostering good health in children than was undertaking more broadly based measures intended to rectify economic inequality. This assumption remains very much in evidence in the current period, with an emphasis on 'parenting skills' having become a well-established and accepted strategy for dealing with the exigencies of poverty (see Edwards 1995).

In her analysis of one woman's experience of psychosurgery in contemporary (1970s) Britain, Hyatt draws upon the history of the maternal and infant welfare programmes of the earlier part of this century to demonstrate a certain continuity in treatments of poverty, which have addressed it primarily as a *medical problem of individuals*, rather than as a systemic problem inherent in the political economy of industrial capitalism. As current worries over supposed drastic declines in population sizes in Europe have again arisen to bedevil policy-makers and politicians (see Spector 1998, for example), we are likely to see a revival of maternalist initiatives which conflate notions about mothering with concerns about the social and biological well-being of people and citizens at the level of populations. Moreover, as economic and material inequality continues to grow within western countries, the present potential for an increase in projects that claim to treat and to avert social disorder through biomedical interventions appears to be great.

While Hyatt shows how, in the earlier part of this century, policy-makers gave disproportionate attention to re-shaping the practices of poor mothers at the expense of addressing the causes and consequences of structural inequality, Kaufert (Chapter 7) illustrates how recent health policy initiatives have completely failed to take into account the conditions, both social and biological, that affect women's health. Even while women have been accorded an important role as the reproducers of populations, their particular health needs as individuals in their own rights have often remained invisible.

The end result is that initiatives that purport to focus on the health of 'the population' are, in fact, initiatives that address primarily those conditions affecting only one group of people: that is, middle-class *men*. In other words, despite the apparent universality and inclusivity of the term 'population', Kaufert shows how current health policy initiatives do not, in fact, take into account the needs of women, the poor, ethnic and racial minorities and others. Health policies are, therefore, shown to be driven by goals and outcomes determined by data that actually omit or obscure large sectors of the population.

Kaufert's chapter also illustrates the very important point that in looking at the relationship between health and gender, it is not enough to just 'add women and stir'. The question her chapter sets out to answer is not, 'how would health policy change if women were included'? but, rather, 'how did the notion of the population become defined through policy discourse in such a way that certain groups, such as women, got left out in the first place?' This is a critical question because it points to the ways in which apparently straightforward and seemingly neutral concepts, such as 'the health of the population', are actually laden with underlying social and cultural biases about which groups' needs are really paramount. Those communities who are most in need of a comprehensive health policy, such as the poor, the elderly, women, and the unemployed to name but a few, end up being the ones who are actually least likely to benefit from a range of new governmental health initiatives.

Littlewood (Chapter 8) tackles a very different set of questions in his chapter. As he points out, there is voluminous research to show that women report much greater incidences of mental illness and of psychological distress than do men; moreover, women are more frequently diagnosed by medical practitioners as suffering from psychological problems and illnesses. Yet, Littlewood emphasises that there is little evidence of any organic or biological cause for these higher rates of mental illness. Rather, Littlewood suggests that women presenting with certain symptoms (such as sleeplessness, for example) are much more likely than men to be diagnosed by practitioners as suffering from mental rather than physical disorders; moreover, the pharmaceutical industry has specifically targeted women as the primary consumers of psychotropic drugs, such as tranquillisers. Littlewood points out that even in western countries, where there are generally much higher degrees of parity between men and women than in the developing world, women are still seen as more constrained by social conventions from taking action to resolve their problems than are men. Therefore, they become the 'perfect' target for medicalised interventions: that is, the passive patient.

In his chapter, Littlewood deals with a particular medical and psychological event disproportionately attributed to women in western societies: the taking of drug overdoses or 'parasuicide' – that is, a suicide attempt undertaken in such a manner, particularly through the misuse of prescription drugs, that it is unlikely to be fatal. Littlewood compares western women's experiences of parasuicide with comparable episodes of apparent psychological disturbances in non-western societies, such as 'spirit possession' and 'wild man possession'.

What links the experience of the British 'housewife' who takes an overdose of sleeping pills and who is rushed to hospital and revived to, say, that of the possessed 'wild man' of highland New Guinea is, in Littlewood's view, the fact of their similar feelings of powerlessness or social dislocation. Acts of self-injury or behaviours that set them apart from others dramatically demonstrate to sufferers' respective communities the degree to which they feel left out, beset, and 'unable to cope'. The desired result of such actions, Littlewood suggests, is not death nor permanent social exclusion but is, rather, a re-acceptance into the community. Through this innovative hypothesis, Littlewood points out that most societies have mechanisms that allow for periodic deviation from psychological 'norms' and that these so-called 'norms' of behaviour themselves vary widely throughout different societies making the diagnosis of mental illnesses particularly susceptible to cultural influences.

Diagnoses of mental illness affect other subordinated groups in western societies, in addition to women. Westwood (1992) for example, has discussed the work of the Black Mental Health Group in Leicester, England, an organisation that sought to call attention to the fact that Black people – in this case, men in particular – were disproportionately being diagnosed as suffering from mental illnesses and were consequently over-represented as patients in secure mental hospitals. Consistent with Littlewood's arguments, Westwood also illustrates that, as in the case of women, Black people's distress was also re-interpreted as psychiatric in nature, thereby allowing doctors and others to sidestep relations of power and powerlessness inherent in hierarchical social structures based on race (and gender) in Britain.

Science and culture

Much of the debate between biological science and social science actually centres around the question of programmatic and policy outcomes. Some social scientists have gone so far as to suggest that acknowledging *any* biological absolutes or commonalities among people inevitably leads to oppressive and reductionist social policies. In their excellent discussion of the current scepticism in some circles about the 'truth' of science, Ehrenreich and McIntosh (1997:15) have explained that according to that critical point of view 'Biology is rhetorically yoked to "determinism", a concept that threatens to clip our wings and lay waste to our utopian visions, while culture is viewed as a domain where power relations with other human beings are the only obstacle to freedom'.

Certainly even the most ardent social constructionist among us would be unlikely to advocate disregard for, say, the germ theory of disease transmission. The problem comes when scientific discoveries are then coupled with moral judgments about individual behaviours. That coupling of science and morality makes clear the point that biology cannot be separated from its cultural context. In the throes of the industrial revolution, for example, poor mothers were not bad mothers because of their moral failings; rather, they lacked access to new information about the causes of infant mortality and the economic resources necessary to implement measures to prevent this tragic outcome.

There is an undeniably unsavoury side to the history of scientific studies of health, in which such groups as women, ethnic minorities, homosexuals and colonised peoples were held up as biological 'deviants' in comparison to the 'ideal' physical type who was inevitably male, white, European, heterosexual, and bourgeois (see Urla and Terry 1995, Marks and Worboys 1997). Many historians and anthropologists of science, including Harraway (1989), Horn (1995) and Gould (1981) have unearthed the connections between social agendas (such as the justification of racial and gender inequalities) and scientific research. Their contributions have been critical to our understanding of how science is, in fact, itself a cultural practice, one whose assumptions, categories, hypotheses and conventions are inevitably derived from cultural understandings of the world. Like all cultural practices, science historically has also been influenced by shifting political imperatives and has been used in the service both of upholding hierarchies based on race and gender and of challenging such hierarchies.

This idea that science is subject to social pressures does not mean, however, that we intend to invalidate the many contributions of scientific research. Nor do we intend to suggest that science and culture constantly pose challenges to one another's insights. Rather, we hope to have demonstrated that an understanding of science and culture in tandem, in this case using the notions of 'sex' as a biological concept and 'gender' as a cultural one, is essential for furthering our analyses of men's and women's experiences of health and disease.

References

Aden, A., Omar, M., Omar, H., Hogberg, U., Persson, L. and Wall, S. (1997). Excess female mortality in rural Somalia: is inequality in the household a risk factor? *Social Science and Medicine,* **44**, 709–15.

Davis, D. (1996). The cultural constructions of the premenstrual and menopause syndromes. In *Gender and Health: An International Perspective,* ed. C. Sargent and C. Brettell, pp. 57–86. New Jersey: Prentice Hall.

Edwards, J. (1995). 'Parenting Skills': views of community health and social service providers about the needs of their clients. *Journal of Social Policy,* **24**, 237–59.

Ehrenreich, B. and McIntosh, J. (1997). The new creationism: biology under attack. *The Nation,* June 9, 11–16.

Fuller, T., Edwards, J., Sermsri, S. and Vorakitphokatorn, S. (1993). Gender and health: some Asian evidence. *Journal of Health and Social Behavior,* **34**, 252–71.

Gann, P., Hennekens, C., Ma, J., Longcope, C. and Stampfer, M. (1996). Prospective study of sex hormone levels and risk of prostate cancer. *Journal of the National Cancer Institute,* **88**, 1118–26.

Gould, S.J. (1981). *The Mismeasure of Man.* New York: Norton.

Harraway, D. (1989). *Primate Visions: Gender, Race and Nature in the World of Modern Science.* New York: Routledge.

Hart, N. (1989). Sex, gender and survival: inequalities of life chances between European men and women. In *Health Inequalities in European Countries,* ed. J. Fox, pp. 109–41. Aldershot: Gower.

Holden, C. (1987). Why do women live longer than men? *Science,* **238**, 158–60.

Holloway, M. (1994). Trends in women's health: a global view. *Scientific American,* **271**, 66–73.

Horn, D. (1995). This norm which is not one. In *Deviant Bodies: Critical Perspectives on Difference in Science and Popular Culture,* ed. J. Terry and J. Urla, pp. 109–28. Bloomington: Indiana University Press.

Lorber, J. (1997). *Gender and the Social Construction of Illness*, Thousand Oaks: Sage.

Macintyre, S., Hunt, K. and Sweeting, H. (1996). Gender differences in health: are things really as simple as they seem? *Social Science and Medicine*, **42**, 617–24.

Marks, L. and Worboys, M. (eds). (1997). *Migrants, Minorities and Health*. London: Routledge.

Martin, E. (1987). *The Woman in the Body: A Cultural Analysis of Reproduction*. Boston: Beacon Press.

Spector, M. (1998). The baby bust: a special report; population implosion worries a graying Europe. *The New York Times*, July 10.

United Nations Development Programme (1992). *Human Development Report 1992*. New York: Oxford University Press.

Urla, J. and Terry, J. (1995). Introduction: mapping embodied deviance. In *Deviant bodies: Critical Perspectives on Difference in Science and Popular Culture*, ed. J. Terry and J. Urla, pp. 1–18. Bloomington: Indiana University Press.

Waldron, I. (1983). Sex differences in human mortality: the role of genetic factors. *Social Science and Medicine*, **17**, 321–33.

Walker, P. and Cook, D. (1998). Gender and sex: vive la difference. *American Journal of Physical Anthropology*, **106**, 255–9.

Westwood, S. (1992). Power/knowledge: the politics of transformative research. *Studies in the Education of Adults*, **24**, 191–8.

2

Parental manipulation of postnatal survival and well-being: are parental sex preferences adaptive?

CATHERINE M. HILL AND HELEN L. BALL

> Concerned to know the truth about the incidence of female infanticide in the Swatow region of Imperial China, missionary and naturalist Adele Fielde "took accounts from forty women . . . each over 50 years of age." She found that the women had borne a total of 183 sons and 175 daughters, however, they had destroyed 78 of their daughters.
>
> (Wolf and Huang 1980:231)

Introduction

In many societies people have been reported to kill infants of the 'wrong' sex. For most of these societies the 'wrong' or undesired sex is female. Female infanticide is well known in parts of India and China; it is also recorded in the Americas, Melanesia, many pre-modern European societies, and is thought to occur, or have occurred, in approximately 9% of the world's cultures (Minturn and Stashak 1982). Male infanticide is much less common, but has been reported for 2–3% of societies. Where infanticide is prohibited sex-biased investment in their children by parents is widespread. Sons and daughters, for instance, may experience differential provisioning: many African groups are reported as breastfeeding girls for longer than boys, while Asian, Latin American, and some European cultures seem to favour breastfeeding males for longer (Hrdy 1987:99). In societies with a strong son preference, combined with the technological means for reliable fertility control (e.g. Korea), parents pursue the 'ideal' family format of more sons than daughters via a sequential decision-making process (Park 1983). The sex of the first two offspring determines their likelihood of continuing to reproduce; for instance if two girls are produced parents will keep reproducing until they have at least two sons; however, if the first two offspring are boys, parents will generally terminate their reproduction at this point. In other areas, such as parts of India (Patel 1989 cited in Laland *et al.* 1995), amniocentesis is increasingly used as a means of prenatal sex-determination which is followed by abortion of female foetuses where parents favour sons; in 1987 in the state of Maharashtra about 40 000 abortions of female foetuses were estimated to have been conducted following sex-determination via amniocentesis (Venkatachalam and Srinivasan 1993:30).

In this chapter we take female infanticide – the most extreme example of sex-biased reduction in parental investment – as a starting place for our examination of parental sex preferences. After reviewing some of the literature on infanticide and sex preferences, we present summaries of several studies which investigate parental strategies regarding preferences for a particular sex of offspring. We then précis our own cross-cultural research regarding the proximate cues that precipitate female infanticide, and parental sex preferences generally. Our aim is to show whether such preferences can be considered to be adaptive in the evolutionary sense.

Background: human infanticidal practices

> . . . an undesired girl is stifled by the mother, father, or grandmother, as soon as her sex is known. A neighbor of one of my Bible-women bore six daughters successively, and smothered five of them. When the sixth came, she said it was always the same girl coming back, and she would no longer endure her. She wanted boys, and would see whether that girl could be deterred from again presenting herself. She cut the child into minute particles, and scattered them over the rice fields.
>
> (Wolf and Huang 1980:231)

There is a substantial literature on differential parental investment and investment allocation by sex in mammals, including humans (e.g. Dickman 1975, Williamson 1978, Scrimshaw 1984, Johansson 1984, 1987, Hrdy and Hausfater 1984, Parmigiani and vom Saal 1994). As noted above, it is well documented that humans regulate parental investment via a continuum of behaviours, from infanticide, through various forms of neglect, to differential allocation of food and health care (e.g. Langer 1974, Dickeman 1975, Scrimshaw 1984, Boswell 1988). Infanticide commonly takes place soon after birth, before the infant has gained the status of a real person in society. In most cases it is one of the parents (primarily the mother), or sometimes the birth attendant, who is responsible for killing the child.

Ethnographers' reports of parental accounts indicate that infants may be killed at birth because they are unlikely to survive or require disproportionate parental care; that is, they are deformed, ill or non-viable (e.g. !Kung, Yanomama, Mundurucu); there are inadequate resources available to rear them, either within the family (e.g. Ayoreo Indians) or within the population (e.g. the Tikopia); they are twins (e.g. Aranda); they are the products of 'abnormal births' and/or considered ill omens (e.g. Bariba); or they are conceived illegitimately (e.g. Aymara) (see Ball and Hill 1996, Hill and Ball 1996). In many of these situations it makes evolutionary sense to kill infants, and as Hrdy (1992) notes, 'children killed are often those whose survival prospects are compromised'. Evolutionary theorists suggest that, in such cases, parents use proximate cues to help them determine when to cut their losses early, kill an infant and invest in the next one (Scrimshaw 1984, Hrdy 1992). This evolutionary (sociobiological) approach assumes that parents act in order to maximise their own reproductive success by favouring those offspring which will, themselves, have high reproductive success – explaining parental investment behaviour in terms of consequences for gene survival and spread within a population. The most adaptive strategy is the one that allows the parents' genes to be carried successfully through to following generations, and this may mean that it is adaptive to invest only in children who are likely to thrive, particularly for a species in which parental investment is high. Thus evolutionary explanations have applied specifically to explain the adaptive significance of the killing of infants who lack paternal support, for whom there are inadequate resources, or whose mother has died. Likewise we recently examined the adaptive significance of the infanticide of twin infants (Ball and Hill 1996) and have argued that 'ill-omens' which are part of the birth circumstances operate as proximate cues that serve to terminate parental investment in situations where it might otherwise be wasted (Hill and Ball 1996).

If, when making investment decisions, parents respond to particular cues, certain biological characteristics pertaining to the neonate will jeopardise its survival chances in an infanticidal culture. These intrinsic factors such as (i) being born with a serious deformity, (ii) being one of twins or other multiple birth, (iii) being born under abnormal circumstances such as by breech presentation or with natal teeth, consistently render infants risky vehicles for parental investment across a wide range of societies, irrespective of environmental, social and cultural constraints (Hill and Ball 1996). Other factors external, or extrinsic, to the child may also influence a parent's decision whether or not to invest in a particular infant (Worthman 1996). When parents use extrinsic cues for investment decision-making the outcome will tend to be dependent on the ecological and cultural environment at the time the decision is made. Examples of such extrinsic cues include illegitimacy, being born under circumstances of inadequate resources, being born too soon after a sibling, and being of the undesired gender (Ball and Hill, unpublished).

When parents terminate investment in offspring on the basis of intrinsic cues they are generally operating in their evolutionary best interests – the infants killed are unlikely to survive or reproduce, and therefore parents' reproductive success will be compromised if investment is wasted upon them (Hill and Ball, unpublished). However, when parents terminate investment on the basis of extrinsic cues the evolutionary argument can have less relevance. In such cases parents are killing intrinsically viable infants. It may be the case that the infant's survival chances are jeopardised by the parent's social or economic circumstances (e.g. has too many children, has too few resources, has no male support). We refer to this as the 'economic constraint scenario'. Alternatively the parents may be unwilling to incur the costs (present or future) of raising this infant (e.g. unwilling to invest in offspring that will incur marriage payments in future). We call this the 'economic choice scenario'. The parents are cast as rational actors favouring one infant over another according to the social or economic costs of rearing. Consideration of these different approaches (which are not mutually exclusive) lead us to suggest that while decisions to terminate investment based on offspring-intrinsic features are likely to be adaptive, infanticide decisions based on offspring-extrinsic features may be adaptive, non-adaptive, or maladaptive in their evolutionary consequences.

Sex preferences

> It is difficult to say when infanticide died out in Kangra, or even whether it is entirely extinct today . . . although infanticide is now undoubtedly extremely rare, infant mortality remains higher for girls than for boys. It would certainly be an exaggeration to describe this as the result of callous neglect, but it is the case that boys are cherished and pampered and have first call on the household's resources, while the attention devoted to a girl is much more perfunctory. I was also struck, for example, by the contrast between two Rajput twins of the house-cluster in which I lived. The boy was chubby and well-dressed; his sister decidedly scraggy and her clothes generally in tatters. When they were babies their mother did not have enough milk to feed both her children, and it was the boy who was breastfed.
>
> (Parry 1979:28)

In this context we can consider sex to be intrinsic to an infant, while its gender says something about the role that an individual will play within its society. Emic and etic explanations for overt son preference have included the need for male strength and economic contribution through work; their importance as warriors; their importance in rituals; and the high cost of dowries (Hrdy 1987:100). However, the sociobiological argument that sons and daughters are likely to provide parents with differing levels of reproductive returns, as explicated by Trivers and Willard (1973) and Hamilton (1967), is unlikely to be recognised by either ethnographers or their informants. The Trivers–Willard hypothesis, in its broadest terms, proposes that if one sex has greater variance in reproductive success than the other (meaning that the upper limit of offspring production can be far greater for one sex than the other), and offsprings' reproductive success is influenced by parental investment, then parents should preferentially invest in the sex with greater variance. In many mammal species, including humans, males have greater reproductive variance than females (Trivers 1972).

Applying the above paradigm to humans, one would expect that parents with resources to invest will favour offspring of the sex with the greatest potential reproductive success, that is, normally, males. Humans, however, are a special case. In explaining human behaviour we must take into account a decision-making process which takes place within the constraints of ideas and values that are culturally based. Viewed from the emic/etic perspective, parents favour offspring of the sex whose co-operation enhances the lineage's short to medium term security, for example access to food or mates, protection of offspring. Likewise parents may kill or reduce their investment in offspring that are likely to be a liability to the family. Thus parents anticipate their infant's socially constructed gender-based role and use this projection to identify infants as poor vehicles for parental investment under specific socio-cultural or environmental circumstances (Hill and Ball, unpublished). We present several case studies below to illustrate how extrinsic cues in the parents' immediate circumstances interact with gender to influence investment decisions.

Ecology and endogamy – the Netsilik Eskimos

Among the Canadian Eskimos of the central Arctic, the Netsilik and Copper Eskimos have a well documented history of female infanticide, while their neighbours the Iglulik Eskimos were reportedly non-infanticidal. The Copper and Netsilik groups experience particularly harsh ecological pressure with long winters and very short open water seasons while the Iglulik experience a longer season of open water (Riches 1974). Balicki (1967) has argued that variations in Eskimo infanticide decisions reflect both temporary ecological crises and long-term perceptions of the difficulties of procuring food. For the Netsilik Eskimos, however, Riches (1974) suggests that infanticidal decisions are further influenced by cultural traits which interact with the prevailing environmental constraints. These include a strong preference for kindred endogamy in the form of first cousin marriage, and the practise of infant betrothal. Parents' investment decisions regarding infant daughters could, we argue, have been influenced by the presence of extrinsic cues. Based on ethnographers' reports, the cue in this instance was whether there was a male first cousin to whom she could be betrothed. If there were not, parents would kill their infant daughters. This economically motivated parental strategy had long-term survival value, as every surviving girl would be sure to have a husband to provide food (an already scarce resource) and would not be a liability to her parents' household economy. The evolutionary consequences meant that parental investment was directed towards daughters who were likely to reproduce, while daughters who were not likely to enhance parental reproductive success (i.e. not betrothed) were eliminated at birth. The inclusive fitness of parents was also enhanced via kindred endogamy. In this example therefore, the infanticidal decision based on the present infant's gender plus an extrinsic cue (presence of an unbetrothed male cousin), was in the lineage's best interests in evolutionary terms, and can thus be considered an adaptive outcome.

Prestige and sons among Chinese peasants

The historical example of female infanticide among Chinese peasants illustrates a decision based on extrinsic cues with evolutionarily maladaptive consequences. Wolf and Huang (1980) describe how males were valued as descendants while females were despised as 'useless things who only grow up and marry out of the family'. Male-biased parental investment, which was based upon parental decisions regarding the economic well-being and prestige of the family (in its immediate lifetime), has resulted in a chronic shortage of women of marriageable age (Wolf and Huang 1980). The extrinsic cue in this instance was the parents' current socioeconomic position, and their anticipation of how an offspring might raise or lower this. If parents thought a daughter would be a liability they killed her. Parental investment was directed to sons with dire consequences for the reproductive fitness of the parents; with no mates available to them the sons of subsequent generations failed to reproduce, leaving their parents with no grandchildren. In pursuit of prestige therefore, Chinese peasants killed their infant daughters to the detriment of their own reproductive success, in evolutionary terms. There is some evidence that female infanticide and neglect of females continues to the present day (*The dying rooms*, Lauderdale Productions 1995).

Daughter rejection and maternal age in Papua New Guinea

> Infanticide was widespread in aboriginal New Guinea . . . [and] female infanticide was quite common . . . Their explanation is couched in very practical terms: "girls don't stay with us when they grow up. They marry and go to other places. They don't become warriors, and they don't stay to look after us in our old age"
>
> (Langness 1981:14–15)

Schiefenhovel (1989) investigated the proximate causes for infanticide among the Eipo, a society of horticulturists in the west New Guinea mountains. This society is characterised as unsegmented and acephalic with patrilineal exogamous clans, virilocal residence, and (mostly) monogamous marriages. Schiefenhovel reported that women monitored the sex of their offspring. In the first half of their reproductive lives they allowed more girls to survive than they did in the second. As women who had no or few sons approached menopause they increasingly killed their newborn daughters (p.176). The consensus of the community was that a mother had the right to reject or accept her newborn child. Twenty-three cases of infanticide occurred during Schiefenhovel's study, some of which were witnessed by him or his wife, and of these 23 they were able to obtain comments regarding motives from the women in 16 cases. In none of the cases was it reported that a newborn was killed because it was male; however, in 3 cases mothers had explicit plans to keep a son and reject a daughter. In attempting to ensure that they have offspring of both sexes, mothers were practising a strategy that potentially reduced their reproductive success (killing viable daughters) but enhanced their short-term economic security, by having sons. Thus the interaction of two extrinsic cues – the necessity of son-support for aging mothers, plus the advancing age of a mother in conjunction with the sex of the present infant – precipitated female infanticide. Because some girls were allowed to live, however, this strategy avoids the extreme maladaptive costs seen in the Chinese peasants.

Female infanticide and neglect on the Indian sub-continent

> The boys are better clad, and when ill are more carefully tended. They are allowed to eat their fill before anything is given to the girl. In poor families, when there is not enough for all, it is invariably the girls who suffer.
>
> (Gait 1911, cited in Poffenberger 1981:79)

On the Indian sub-continent the preference for sons is widespread, and female infanticide and female neglect are common. Investigating the incidence of female infanticide in Tamil Nadu, Venkatachalam and Srinivasan (1993) found that 9% ($N = 111$) of their respondents (all women) admitted to killing a daughter within the previous two years, and that 38% ($N = 476$) claimed they would have to commit female infanticide if they bore more than one daughter. The use of prenatal sex-determination followed by female foeticide (sometimes as late as the 7th month of pregnancy) was reported by 7% of respondents ($N = 89$). These findings are partly explained by the belief expressed by 69% of respondents that a female child is a liability, with the expense of the obligatory ceremonies that would have to be performed on her behalf given as the main reason for female infanticide. The fact that a daughter would leave her parents' household upon marriage (Poffenberger 1981), and that parents would thus gain little from investment in her is expressed in the saying 'bringing up a girl is like watering the neighbour's plant' (Venkatachalam and Srinivasan 1993:31). If they are not killed before or at birth, son preference is so prevalent that girls are commonly denied adequate food, clothing, education, and access to medical care (Venkatachalam and Srinivasan 1993, Poffenberger 1981, Rousham, Chapter 3). While there was no evidence of direct infanticide of females, data from Rajpur in Northern India reveal that the majority of parents believe that having a daughter is economically too costly (Poffenberger 1981). One father is quoted as saying 'There is no benefit in having a daughter. She goes to another family when she is 15 or 16 years old' (Poffenberger 1981:82). Consequently, daughters become the victims of neglect, and while boys are believed to need special attention from their parents, girls are considered more resilient and in need of less care. The outcome would appear to be neither adaptive nor maladaptive because this strategy does not obviously affect the reproductive success of parents. The proximate extrinsic cues in this example – anticipation of economic security, plus at least one existing daughter – interact with the sex of the current infant to determine whether or not it will be killed or neglected.

Miller developed a sociocultural and economic model to explain parental sex preferences in India, and tested this model using data from the neighbouring countries of Pakistan and Bangladesh (Miller 1984). Based upon Indian ethnographic and census data on the utility of females in agriculture, and on ethnographic information on marriage practices and the costs of dowry versus bridewealth systems, she predicted that females in Pakistan would be little valued due to their low economic utility and liability of their dowries, leading to a strong son preference. Meanwhile in Bangladesh, where female labour was highly valued, and marriage costs for daughters were low (or even income-producing) Miller predicted there would be neither son preference nor daughter neglect (1984:119). Her findings confirmed that the economic importance of females in Bangladeshi households was greater than that of females in Pakistani households – with female economic contribution being more important than marriage payments (p.122). She concludes that, in spite of economic and cultural reasons for son preference in both Pakistan and Bangladesh, daughters have better survival chances in the latter because their role in food processing has, in Bangladesh, been traditionally large and important (Miller 1984:122). Rates of female infanticide correlate with extrinsic cues centred on immediate economic costs and benefits of daughters.

Preferential investment in daughters – the Mukogodo of Kenya

The Mukogodo are a pastoralist group living in Laikipia District, Kenya. Within this population the sex ratios at birth appear to be normal but within the 0–4 yrs age cohort the sex ratio is skewed in favour of girls. Cronk suggested that this skewed sex ratio arises as a consequence of differential parental solicitude favouring girls, combined with a higher than normal neonatal mortality rate for boys (Cronk 1991a). Parents invest more in the health care of their daughters than of their sons. Interestingly, 'Although their behavior and the childhood sex ratio suggest a strong tendency to favor daughters, many mothers express a preference for sons, and their overall stated sex preference is roughly equal' (Cronk 1991b: 399).

Why should parents favour daughters, practically if not 'culturally', among the Mukogodo? According to Cronk (1991a) there is no evidence to support the hypothesis that daughters enhance their family's economic standing; instead it appears that the reason the Mukogodo favour their girls is because daughters have better marriage prospects than do sons. The marriage system practised by the Mukogodo is generally one of hypergynous marriages (women marrying up the social hierachy); thus, because most people are of low social status and impoverished, all daughters marry but some sons will not marry. In this situation sons represent wasted parental investment while daughters have greater reproductive potential. Enhanced investment in daughters in this society is, therefore, evolutionarily adaptive, and is as predicted under the Trivers–Willard hypothesis.

Differential parental investment in the Ecuadorian Highlands

Throughout the Ecuadorian Highlands many people believe that certain qualities pertaining to sexuality and aggression are transmitted from mother to child during breastfeeding. Whilst these characteristics are seen as 'ideal' and 'appropriate' with respect to men, they are inappropriate traits in women. If a mother nurses her daughter for longer than one year, then when that daughter reaches sexual maturity it is thought that she will become *grosa*, i.e. a 'vulgar woman'. Thus throughout many of the Highland regions girls are reported as being nursed for a significantly shorter time than are their brothers. McKee (1977, 1982 in McKee 1984) recorded a mean average weaning age for girls of 10.93 months as compared with the mean average weaning age for boys of 20.27 months.

National-level statistics on childhood mortality in the Highland Provinces of Ecuador, taken over several different years, indicated that the mortality rates for female children increased significantly during the second year of life. The timing of this increase in female infant mortality would appear to coincide with the age at which they are weaned. Although mean average weaning ages for girls and boys differ across the Highland Provinces, girls are always weaned earlier than boys and it appears that this may, at least, contribute to their overall higher mortality (McKee 1984). The proximate cues determining a mother's degree of lactational investment in any one child appears to be the interaction of the child's sex with cultural beliefs regarding the effects of lengthy breastfeeding, which facilitate and encourage the mother's withdrawal of that investment. If girls are a liability to their parents in some way (e.g. economic or social), this strategy serves both to reduce parental investment in girls and shorten the inter-birth interval following the birth of girls but not boys. Differential investment operates in response to cultural explanations but it is unclear whether there are any evolutionary advantages to this strategy.

Proximate cues and sex preferences: a cross-cultural test

In a recent paper (Ball and Hill unpublished) we tested the 'economic choice' explanation. We hypothesised that female infanticide will be practised in those societies where infanticide is permissible and female offspring are a disproportionate economic liability to their parents compared with male offspring (where 'economic' pertains to both household labour and material wealth). We also hypothesised that neglect or abandonment would be more prevalent among offspring of the sex which is the greatest liability.

We used Murdock and White's (1969) 'Standard cross cultural sample' of 186 societies and examined all obtainable ethnographies for information on infanticide and parental sex preferences. The ethnographic literature suggested that there are several scenarios under which parents prefer sons over daughters. To determine whether these could be generalised cross-culturally we proposed and tested several predictions involving both infanticide and differential investment. For instance, offspring of the sex that makes a large contribution to the household economy will be cheaper to rear than non-contributing offspring. Therefore, we predicted that, in those societies where infanticide is permissible, female infants will be killed where female contribution to household resources is low, and male offspring will be preferred if female contribution is low. Likewise, we made predictions regarding the exchange of material or labour resources for females upon marriage, the payment of dowries upon marriage of daughters (e.g. male infants will be preferred where dowries are obligatory), residence rules for daughters following marriage, and inheritance practices.

We obtained data on the presence/absence of infanticidal practices for 127 of Murdock and White's 186 societies. Data on which infants were reported to be killed, presence/absence of parental sex preferences, and which offspring were reported to be preferred, were entered into a spreadsheet. These data were augmented with coded data from Murdock's *Atlas of World Cultures* (1981) for predominant subsistence activity, mode of marriage, marital residence, and inheritance of moveable property. Additionally coded data on the proportion of female contribution to subsistence obtained from Schlegel and Barry (1986), and on children's contribution to childcare (Barry and Paxson 1971) were included.

Data were tabulated as a series of 2 × 2 contingency tables in order to test the predictions given above. Statistical analyses were carried out using Chi Square and Fisher's Exact Test. Significance was accepted at the 5% level (i.e. $p < 0.05$). Infanticide was reported for 42% ($N = 53$) of those societies examined. The ethnographies of 12 societies explicitly indicated that female infanticide was practised, while in 26 societies, although infanticide was reported, girls were not included as a specific category to be killed. For the remaining 15 societies it was unclear as to which infants were killed. Specific information on parental sex preferences was available from ethnographies for 115 of the 127 societies examined (90.6%). Forty-five of these (39%) preferred sons, 17 (14.8%) preferred daughters, and 53 (46.1%) were reported not to favour offspring of one sex over the other.

Our results indicated that the proximate variables associated with female-biased infanticide, and probable discrimination against females, were consistent with the hypothesis that parental investment decisions are influenced by factors pertaining to economic choice, i.e. the proportional contribution made by females to household economy, the receipt of bride price, and (for sex preferences) daughters' residency patterns after marriage. We concluded that, cross-culturally, parents tended to practice early termination of investment in females where daughters were an economic liability, making little contribution to family resources via work or marriage payments (Ball and Hill unpublished).

We acknowledge that there are some problems inherent in these kinds of data. Firstly, the information within the database has come from many different ethnographic sources, therefore, details of different societies' infanticidal practices or sex preferences do not necessarily conform to specific, predetermined categories, which may introduce errors. Secondly, the information has passed through several filters, for example reported cases of infanticide may have been categorised differently by various ethnographers. Thirdly, the limitations of these data, and our lumping of them into broad categories, combined with the type of analysis we attempted, means that much of the detail originally present in the ethnographic data was lost. Fourthly, Cronk has presented evidence which suggests that what people say with respect to parental sex-preferences may not be what they actually do (Cronk 1991a). Finally we may have under-represented societies where the most detailed ethnographies are not in English. However, use of cross-cultural data on this scale minimises the biases which result from reliance on a small number of ethnographic sources.

General discussion

Our findings support the results of Hewlett (1988 cited in Sieff 1990) who conducted a comparative study on the relationship between sex-biased parental investment and relative contributions of sons and daughters in 10 populations of tropical hunter–gatherers. Hewlett concluded that male children are favoured in societies where males contribute more to subsistence. Conversely we have also found examples, such as the Gwembe Tonga of Zambia, where female contribution to subsistence activities is very high (71%), and where both mothers and fathers state a preference for daughters. Furthermore the extremely high mortality rate reported for male infants in this society is suspected to be a consequence of selective neglect, and perhaps infanticide (Clark *et al.* 1995).

Cronk (1991b) recommends that researchers look for female biases in parental investment wherever females are more economically productive than males, attract bride-wealth payments, are used in direct exchange mating systems, or where fathers, sons and brothers are the main competitors for the resources required for mating (p.408). Our results in this regard indicate that the potential for significant economic gain associated with female offspring (labour or bride-wealth) is the predominant circumstance under which parental investment in female infants is unlikely to be terminated. This finding complements the conclusion of Schlegel and Barry (1986) that 'when women contribute substantially to the food supply, they are perceived less as objects for male sexual and reproductive needs and more as persons in their own right. High contribution does not in itself lead to high status, but it may lead to the perception of women as more self-sufficient and less manipulable than they are perceived as being in low contribution societies' (p.149). In other words, such circumstances prevent daughters from being considered liabilities.

Parental investment in a given offspring always involves a trade-off for parents between subsistence needs and reproductive effort, and between investment in present and future offspring. Where a given child has the potential to contribute to subsistence needs of the parents or siblings, the child in question will be less vulnerable to the withdrawal of parental investment in early life (be it by infanticide or reduced investment as the less-favoured sex). For instance Hrdy and Judge (1993) present an example from peasants in Bangladesh where 'a son will become a net producer of calories by the age of 10–13. By age 15, his cumulative lifetime production will have completely repaid his parents for what it cost to rear him. By age 21, he will have paid for himself and one sister. In contrast, even though a daughter begins to work hard at an early age, she will never repay her parents before she marries and leaves to join another family's patriline' (p.22).

If parents are using extrinsic cues, generally based on assessment of lifetime economic liability or worth, to make sex-biased infanticide and investment decisions, can we predict what evolutionary consequences might ensue? To answer this question we need to examine not only issues surrounding investment in the current infant, but implications for future infants as well. Here the intersection of the economic and evolutionary perspectives might benefit from a framework which examines how cultural traditions and practices affect evolutionary outcomes. For instance, it has been demonstrated (Harada 1989 cited in Laland *et al.* 1995) that female infanticide is likely to spread in societies in which paternal influence in cultural transmission is strong (e.g. Copper Eskimo), and that male infanticide (or neglect) is likely to spread in societies where the mother is more influential in cultural transmission (e.g. Gwembe Tonga, Clark *et al.* 1995). Furthermore, influential members of a population may propagate preferences and cultural traditions that ensure their own success, which may be shared by others in the population even if they do not enhance the reproductive success of the majority (e.g. north India).

On considering the cross-cultural range of infanticidal practices it is clear that infanticide triggered by infant-intrinsic cues (based on particular biological features of an infant which tend to reduce its survival chances) should be adaptive. For instance many cultures practice infanticide of 'deformed' infants (see Ball and Hill 1996). We suggest that we would not expect to see a sex-bias in the operation of these proximate cues and that arguments based on evolutionary theory do not provide useful explanations for sex-biased infanticide in humans. Rather, we argue, female infanticide and female neglect result from decisions carried out in response to a wide array of cultural cues pertaining to economic choice. Adaptiveness in these situations is unlikely to be predictable, and will vary according to time and place. Thus parental investment decisions relating to factors associated with the economic choice scenario will serve to promote parents' own economic interests. While such actions do not necessarily maximise parental reproductive success, as long as some offspring survive, and those offspring follow similar parental strategies (or adhere to the same cultural rules), practises, such as infanticide or neglect, for such offspring-extrinsic reasons will be maintained regardless of the evolutionary consequences.

References

Balikci, A. (1967). Female infanticide on the Arctic coast. *Man: The Journal of the Royal Anthropological Institute*, **2**, 615–25.

Ball, H.L. and Hill, C.M. (1996). Re-evaluating twin infanticide. *Current Anthropology*, **37**, 856–63.

Barry, H. and Paxson, L.M. (1971). Infancy and early childhood. Cross-cultural codes 2. *Ethnology*, **10**, 466–508.

Boswell, J. (1988). *The Kindness of Strangers: The Abandonment of Children in Western Europe from Late Antiquity to the Renaissance.* New York: Pantheon.

Clark, S. E., Colson, J. L. and Scudder, T. (1995). Ten thousand Tonga: a longitudinal anthropological study from Southern Zambia. *Population Studies*, **49**, 91–109.

Cronk, L. (1991a). Intention versus behaviour in parental sex preferences among the Mukogodo of Kenya. *Journal of Biosocial Science*, **23**, 229–40.

Cronk, L. (1991b). Preferential parental investment in daughters over sons. *Human Nature*, **2**, 387–417.

Dickeman, M. (1975). Demographic consequences of infanticide in man. *The Annual Review of Ecology and Systematics*, **6**, 107–37.

Hamilton, W. D. (1967). Extraordinary sex ratios. *Science*, **156**, 477–88.

Hill, C. M. and Ball, H. L. (1996). Abnormal births and other 'ill-omens' – The adaptive case for infanticide. *Human Nature,* **7**, 381–401.

Hrdy, S. B. (1987). Sex-biased parental investment among primates and other mammals: a critical evaluation of the Trivers–Willard hypothesis. In *Child Abuse and Neglect: The Non-Human Primate Data*, ed. R. Gelles and J. Lancaster, pp. 91–147. New York: Aldine.

Hrdy, S. B. (1992). Fitness tradeoffs in the history and evolution of delegated mothering with special reference to wet-nursing, abandonment, and infanticide. *Ethology and Sociobiology,* **13**, 409–42.

Hrdy, S. B. and Hausfater, G. (1984). Comparative and evolutionary perspectives on infanticide: introduction and overview. In *Infanticide: Comparative and Evolutionary Perspectives*, ed. G. Hausfater and S. B. Hrdy, xiii–xxxv. New York: Aldine.

Hrdy, S. B. and Judge, D. S. (1993). Darwin and the puzzle of primogeniture: an essay on biases in parental investment. *Human Nature,* **4**, 1–46.

Johansson, S. R. (1984). Deferred infanticide: excess female mortality during childhood. In *Infanticide: Comparative and Evolutionary Perspectives*, ed. G. Hausfater and S. B. Hrdy, pp. 463–86. New York: Aldine.

Johansson, S. R. (1987). Neglect, abuse and avoidable death: parental investment and the mortality of infants and children in the European Tradition. In *Child Abuse and Neglect: Biosocial Dimensions*, ed. R. Gelles and V. Lancaster, pp. 57–93. New York: Aldine.

Laland, K. N., Kumm, J. and Feldman, M. W. (1995). Gene–culture coevolutionary theory: a test case. *Current Anthropology,* **36**, 131–56.

Langer, W. L. (1974). Infanticide: an historical survey. *History of Childhood Quarterly,* **1**, 353–66.

Langness, L. (1981). Child abuse and cultural values: the case of New Guinea. In *Child Abuse and Neglect: Cross Cultural Perspectives*, ed. J. E. Korbin, pp. 13–34. Berkeley: University of California Press.

Lauderdale Productions (1995). *The Dying Rooms.* Shown as part of Channel 4's *Secret Asia* series.

McKee, L. (1984). Sex differentials in survivorship and customary treatment of infants and children. *Medical Anthropology,* **8**, 91–108.

Miller, B. (1984). Daughter neglect, women's work, and marriage: Pakistan and Bangladesh compared. *Medical Anthropology,* **8**, 109–26.

Minturn, L. and Stashak, J. (1982). Infanticide as a terminal abortion procedure. *Behavior Science Research,* **17**, 70–90.

Murdock, G.P. (1981). *Atlas of World Cultures.* Pittsburgh: University of Pittsburgh Press.

Murdock, G. P. and White, D. R. (1969). Standard cross cultural sample. *Ethnology,* **8**, 329–69.

Park, C. B. (1983). Preferences for sons, family size, and sex ratio: an empirical study in Korea. *Demography,* **20**, 333–52.

Parmigiani, S. and vom Saal, F.S. eds. (1994). *Infanticide and Parental Care.* Switzerland: Harwood Academic Publishers.

Parry, J. P. (1979). *Caste and Kinship in Kangra.* London: Routledge and Kegan Paul.

Poffenberger, T. (1981). Child rearing and social structure in Rural India: toward a cross-cultural definition of child abuse and neglect. In *Child Abuse and Neglect: Cross Cultural Perspectives,* ed. J. E. Korbin, pp. 71–95. Berkeley: University of California Press.

Riches, D. (1974). The Netsilik Eskimo: a special case of selective female infanticide. *Ethnology,* **13**, 351–62.

Schiefenhovel, W. (1989). Reproduction and sex-ratio manipulation through preferential female infanticide among the Eipo, in the Highlands of West New Guinea. In *The Sociobiology of Sexual and Reproductive Strategies,* ed. A. Rasa, C. Vogel and E. Voland, pp. 170–93. London: Chapman and Hall.

Schlegel, A. and Barry, H. (1986). The cultural consequences of female contribution to subsistence. *American Anthropologist,* **88**, 142–50.

Scrimshaw, S. C. M. (1984). Infanticide in human populations: societal and individual concerns. In *Infanticide: Comparative and Evolutionary Perspectives,* ed. G. Hausfater and S. B. Hrdy, pp. 170–93. New York: Aldine.

Sieff, D. (1990). Explaining biased sex ratios in human populations. *Current Anthropology,* **31**, 25–48.

Trivers, R. L. (1972). Parental investment and sexual selection. In *Sexual Selection and the Descent of Man,* ed. B. Campbell, pp. 136–79. Chicago: Aldine.

Trivers, R. L. and Willard, D. E. (1973). Natural selection and parental ability to vary the sex-ratio of offspring. *Science,* **179**, 90–2.

Venkatachalam, R. and Srinivasan, V. (1993). *Female Infanticide.* New Delhi: Har-Anand Publications.

Williamson, L. (1978). Infanticide: an anthropological analysis. In *Infanticide and the Value of Life,* ed. M. Kohl, pp. 61–75. Buffalo, New York: Prometheus Books.

Wolf, A. P. and Huang, C. S. (1980). *Marriage and Adoption in China.* Palo Alto, California: Stanford University Press.

Worthman, C. (1996). Biosocial determinants of sex ratios: survivorship, selection and socialization in the early environment. In *Long-Term Consequences of Early Environment, Growth, Development, and the Lifespan Developmental Perspective,* ed. C. J. K. Henry and S. J. Ulijaszek, pp. 44–68. Cambridge: Cambridge University Press.

3

Gender bias in South Asia: effects on child growth and nutritional status

EMILY K. ROUSHAM

Introduction: biological differences in male and female child mortality

There is a consistent phenomenon in human reproduction whereby more males are born than females. The sex ratio at birth is approximately 106 in almost all populations, that is, 106 males are born for every 100 females (Coale 1991). This excess of male births is counterbalanced by higher mortality rates among males at all stages of postnatal life.

The differential mortality of males and females has been attributed to an inherent biological vulnerability of the male sex. One explanation for excess male mortality is the possession of only one X chromosome. In addition to having only single copies of normal genes, males have a greater risk of expressing X-linked recessive disorders than females. In fact, the contribution of X-linked disorders to male infant mortality is thought to be small (Waldron 1983), but the general effect of having only one X chromosome is less well understood. Another explanation for sex differences in mortality is the immaturity of males relative to females at all stages of development. By four months of gestation, skeletal development is three weeks more advanced in the female foetus compared with the male and, at birth, girls are 4–6 weeks more mature than boys. Girls appear to be more mature physiologically in several organ systems and this is thought to contribute to their greater survival at birth (Tanner 1989). For example, males have much greater rates of respiratory distress syndrome in the neonatal period, which in turn arises from the immaturity of the male lungs (Waldron 1983).

Males also have higher death rates from infectious disease during the first year of life and this is thought to arise from lower immune resistance (Waldron 1987).

In most countries throughout the world, the greater mortality rates of males means that the sex ratio decreases across the age span from 106 at birth to a value much closer to 100 in adulthood. There are, however, some notable exceptions to this trend. In some developing countries, female infants and children have higher mortality than males resulting in a sex ratio of adult males to females much greater than one (Coale 1991). This inverse mortality pattern, with females showing higher mortality than males, is most pronounced in the geographical region that extends from Southern Asia, through Western Asia to Northern Africa (Waldron 1987). In general, these populations are associated with strong patriarchal family structures and a cultural preference for male children. Over the last two decades, therefore, it has been increasingly recognised that socio-cultural influences on gender differences in mortality may be superimposed on the biological profile of male and female mortality.

The rest of this chapter is concerned with gender bias in South Asia. This is one of several regions of the world where female children show far greater mortality than expected. Increased female mortality has been attributed to a strong socio-cultural preference for sons which is thought to be translated into differential treatment and care of boys and girls, particularly in infancy and early childhood.

Gender bias in South Asia

Demographic surveys in South Asian countries have revealed male survival rates which are far greater than expected when compared with male mortality rates in developed countries. In the 1991 census, the ratio of adult males to females in India was 107 compared with an expected ratio of 102 (based on a sex ratio at birth of 106; Coale 1991). This figure indicates that many more females than males are dying before adulthood. The same pattern of 'missing' females is observed in population estimates for Pakistan, Bangladesh and Nepal (Coale 1991).

In India, excess female mortality is seen across all age groups, but it is most pronounced among young children from 1 to 5 years of age (Waldron 1987). In southern Bangladesh, D'Souza and Chen (1980) showed that in the neonatal period, male mortality exceeded female mortality consistent with the pattern in other human populations. In the postneonatal period, however, female deaths outnumbered male deaths by as much as 50% in children from 1 to 4 years of age. The excess mortality of female children in Bangladesh has been confirmed by subsequent research in the same region and by the Bangladesh Fertility Survey (Koenig and D'Souza 1986, Majumder *et al.* 1997).

Although national statistics for India indicate that males have better survival chances than females, there is significant geographical variation in male:female mortality ratios. The 1981 census showed excess female mortality to be greatest in the north of India, particularly in Punjab, Haryana, Rajasthan and Bihar (Harriss 1990). This bias towards male survival becomes progressively less significant in the south of India with female mortality being lowest in Tamil Nadu and Kerala. A similar gradient is observed in the sex ratio (Das Gupta 1987, Harriss 1990).

The most common explanation for gender differentials in survival is the cultural preference for sons over daughters. Many parts of South Asia practice exogamy whereby daughters leave their village following marriage and have little further to do with their parents. Daughters require a dowry payment to the groom's household and are therefore seen as presenting a financial drain on their parents. A poor household with several daughters faces the prospect of a lifetime of debts accrued through dowry payments. Within the marital household a daughter-in-law assumes a low social status which may be gradually elevated by bearing children. Her social status is considerably more enhanced, however, if she gives birth to a son.

That son preference is an underlying cause of excess female mortality is supported by the significant correlation between regional differences in male:female mortality rates across India and an index of son preference expressed by adults in these regions (Dyson and Moore 1983). In addition to the cultural explanation for the gender bias in South Asia, some researchers have suggested material and economic causes underlying the poor health and survival of female children (Koenig and D'Souza 1986). Sons are regarded as an economic asset, since they will become the main producers of the family and provide the greatest economic security for their parents in old age. Men receive greater pay for labour than women performing the same tasks, and demand for male labour is higher in most agrarian economies in South Asia. Women are often precluded from ownership of property, which further contributes to their low social and economic status (Kabeer 1991).

Although the sex bias in mortality rates in South Asia is well-established, it has been far more difficult to identify which particular behaviours, practices or attitudes give rise to greater female child mortality. The question arises of whether a sex bias in mortality arises through conscious neglect of daughters, or a benign preference for sons. It is also unclear whether these mortality patterns can be directly linked to the differential morbidity and nutritional status of male and female children (Harriss 1990). To investigate the causes of excess female mortality, researchers have concentrated on two main areas: (1) household feeding patterns and nutrient intake; and (2) rates of illness and treatment-seeking behaviour for male and female children. These two areas will be reviewed in turn.

There is relatively little information concerning gender bias in breastfeeding practices. Harriss (1989) suggests that there is no sex difference in breastfeeding, but in rural Uttar Pradesh female babies were sometimes breastfed for a shorter time and less intensely than male babies (Santow 1995). Other studies have suggested that female infants are breastfed less frequently and for a shorter duration than males (Chatterjee and Lambert 1989), but this has not been widely demonstrated.

It is generally at the time of weaning and the introduction of solid foods that sex discrimination may begin. In particular, the quality and quantity of supplementary foods is thought to vary according to the sex of the child. Das Gupta (1987) reported that ordinary food was divided in an unbiased manner between sons and daughters in north India, but special, more nutritious food often went disproportionately to sons. In Bangladesh, male calorie consumption exceeded female consumption by 16% among children under 5 years and protein intake was, on average, 14% higher in male children (Chen *et al.* 1981). Significantly lower food intakes among females aged from 1 to 4 years have been reported elsewhere in Bangladesh (Abdullah and Wheeler 1985) but there is also one report of greater nutrient intakes for females compared with males (Chaudhury 1984).

The relationship between differential mortality and nutrition is not always clear. In North Indian villages, for example, excess female child mortality has been observed coexisting with apparently unbiased child feeding practices (Harriss 1989). In Nepal, no differences were found in the mechanisms of food distribution or nutrient intake between male and female children, although preferential food distribution was observed for adult males over females (Gittelsohn 1991). This was a population where children were fed first and given the more nutritious foods, hence their dietary intake was adequate for most nutrients. Marriage payments for women were relatively small and mainly comprised personal belongings for the wife. The combination of a high priority given to infant nutrition and health, and low material costs of having daughters may have led to equal feeding of male and female children (Gittelsohn 1991).

Sex differences in growth and anthropometric status provide additional insight into the nutritional status of male and female children. Female children in the Matlab region of southern Bangladesh had markedly higher rates of severe malnutrition (19% vs 6%) from ages 0–4 years (Chen *et al.* 1981). Among rural Indian children, however, no significant sex differences in anthropometric status were observed (Gaur and Singh 1994). In rural Nepal, no sex differences in stunting or wasting were found among children aged 0–49 months (Panter-Brick 1997). Highlighting the importance of geographical variation across the sub-continent is the finding that boys aged from 5 to 7 years in South India were at significantly greater risk of malnutrition than girls (Jeyaseelan and Lakshman 1997).

The second possible cause of differential mortality of the sexes is the provision of health care and medical treatment to boys and girls. Two studies in Bangladesh have reported on sex discrimination in treatment-seeking behaviour. Chen *et al.* (1981) reported that, despite a similar incidence of diarrhoea in male and female children, male children under 5 years of age were brought to the Matlab treatment facility far more frequently than females (treatment rates of 135.6 per 1000 for males cf. 81.9 per 1000 for females). Hossain and Glass (1988) found that drug purchases were greater for male than female children under 5 years of age (incidence rate ratio = 1.71) and this increased to a rate ratio of 2.94 for drugs prescribed by a physician. Among families originating from Uttar Pradesh, boys received more frequent health care than girls. This, as opposed to unequal food distribution, was thought to be the underlying cause of excess female mortality (Okojie 1994). Das Gupta (1987) found that families in northern India spent twice as much on medical care for boys in the first year of life compared with girls.

More detailed community-based studies are required to gain a better understanding of the role of infectious disease and treatment in excess female mortality. Investigation of the onset, severity and parental responses to illnesses of boys and girls as well as the outcome of such incidents would provide valuable insights into the aetiology of sex differentials in mortality.

One of the reasons why feeding patterns and health care have not been found to favour males universally may be because these behaviours are moderated by other household characteristics. Factors which have been shown to affect the survival of boys and girls are: socio-economic status; parental education; birth order and the sex composition of children within the household. Muhuri and Preston (1991) showed that mortality risk in Bangladesh was moderated by the sex composition of children within the family. Girls had a 54% greater risk of mortality than boys overall, but this risk increased to 84% if they had older sisters and dropped to 14% if they did not have sisters. The authors concluded that not all females are discriminated against, but that there is a pattern of 'conscious, selective neglect of individual children' (Muhuri and

Preston 1991). They also showed that parents pursue a child survival strategy aimed at securing some balance between male and female offspring, albeit with a strong bias towards sons.

Das Gupta (1987) also argues that sex discrimination is not applied universally, as shown by highly significant effects of birth order on infant mortality rates. First born infants show the expected mortality rates based on biological differences between the sexes (126.8 deaths per 1000 livebirths for males and 95.7 per 1000 for females aged from 0 to 4 years). For fourth and higher birth order children, however, the rates were 99.3 and 152.7 for males and females respectively. This trend of greater female mortality within higher birth orders was even more pronounced among highly educated women. On the basis of these findings, Das Gupta suggests that the general population trend of increasing levels of education will not be associated with a decline in gender discrimination.

However, the general effect of parental education on gender bias appears to be inconsistent. Some studies have reported no effect of maternal education, or greater female mortality in families with better educated mothers (Das Gupta 1987, Bhuiya and Streatfield 1991). Again, it has been argued that the status of women within society is more fundamental than education alone for the reduction of gender discrimination.

In terms of treatment-seeking behaviour, socio-economic factors or scarcity of resources have also been shown to play a role. Rahman *et al.* (1982) observed approximately equal attendance of boys and girls at a diarrhoeal clinic when including only the families which resided close by. In families living further away, however, the attendance rates for female children were much less than those for males (Rahman *et al.* 1982).

Socio-economic and temporal variation in female undernutrition: a case study from Bangladesh

Previous studies have found significant variations in excess female mortality according to socio-economic status, but few have considered whether socio-economic factors influence sex differences in growth and nutritional status. Furthermore, given that most rural communities in South Asia experience pronounced seasonal changes in food availability and infectious disease prevalence, it is surprising that no previous studies have looked at temporal variation in gender discrimination. The prediction tested in the present study is that gender bias will be greater under adverse economic and environmental conditions. The hypotheses under investigation are firstly, that gender bias in nutritional status will be most significant in the poorest households and secondly, that gender bias will be greatest in the aftermath of natural disasters and during the monsoon season.

These hypotheses were tested using data collected as part of a 16 month longitudinal study of child growth, nutrition and infection in a remote, rural area of Bangladesh (Rousham and Mascie-Taylor 1994). The study area was in the Jamalpur district, northern Bangladesh, which received free primary health care services provided by a non-government organisation. Subsistence was based upon rice and wheat cultivation, with a number of households participating in cash cropping of sugar cane. The study began at a time of widespread famine resulting from the loss of two rice crops. The main rice crop was destroyed by the devastating monsoon floods of the previous year, and the smaller winter harvest was severely reduced by a drought which lasted for the first four months of the study. After these disasters conditions gradually improved over the course of the study.

A total of 1402 children from 2 to 6 years of age were enrolled in the study. Data collection was carried out by local Bangladeshi health workers who came from the surrounding villages and were known to the mothers and children. Height, weight and mid-upper arm circumference were measured every month on household visits. In addition, child morbidity data were collected every two weeks by asking the parent whether the child had experienced any illnesses within the last two weeks, relying on the mother to describe the symptoms. The health workers worked in pairs comprising one male and one female; this was the most acceptable arrangement both for the health workers and the mothers in terms of observing Islamic practices. The height and weight measurements were compared with the WHO/NCHS growth references (a United States data set from the National Center for Health Statistics, recommended as a reference by the World Health Organization) and the Bangladeshi values are expressed as z-scores (standard deviations) from the reference mean. The WHO/NCHS reference data are calculated for boys and girls separately, and take into account the expected biological differences in childhood growth between the sexes.

Sex differences in growth and nutritional status were examined in relation to land ownership, since this is the strongest determinant of socio-economic status in rural society. Landless households relied upon casual employment by wealthier farmers and were usually hired on a daily basis. Employment opportunities were determined by the agricultural cycle and were therefore highly seasonal with the period of greatest unemployment falling during the monsoon season (June–September). Landowning households comprised those who farmed their own land, and a smaller proportion who had professional employment (such as school teachers or government employees) and hired casual labourers to cultivate their land.

The growth of male and female children was examined in relation to socio-economic status (landless or landowner). Analysis of variance was used to test the effect of sex and socio-economic status on height-for-age and weight-for-age as well as testing for statistical interaction between sex and socio-economic status.

Sex differences in growth

Figures 3.1 and 3.2 show the growth of male and female children over the 16 month study in terms of weight-for-age and height-for-age z-scores. The graphs demonstrate that landless female children had significantly worse nutritional status than all other children. Analysis of variance of weight-for-age at baseline revealed a statistically significant interaction between sex and socio-economic status (Table 3.1, $F = 9.09$, $p < 0.003$). This means that there was a synergistic relationship between sex and socio-economic status. That is, the effect of being female and being landless was significantly worse than the effect of either variable individually.

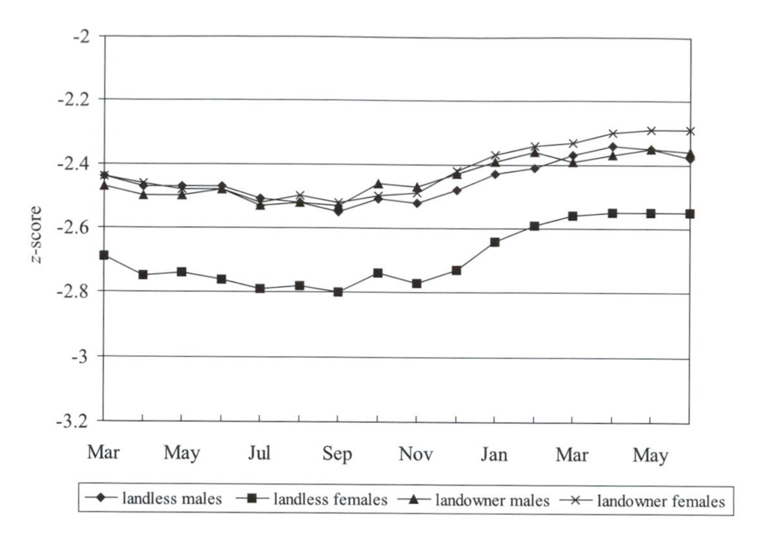

Figure 3.1. Mean weight-for-age of male and female children according to socio-economic status.

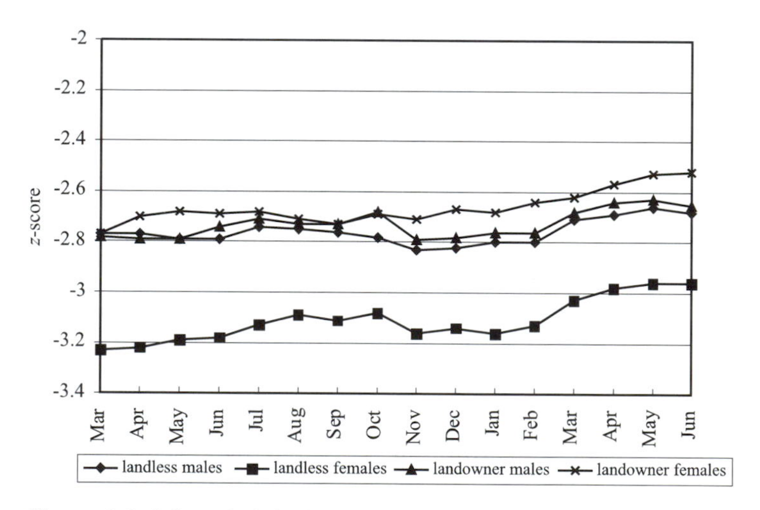

Figure 3.2. Mean height-for-age of male and female children according to socio-economic status.

A highly significant interaction between sex and socio-economic status was also observed for height-for-age *z*-scores at the beginning of the study ($p = 0.002$, see Table 3.1 and Figure 3.2).

Figure 3.3 shows the mean weight-for-height *z*-scores of male

Table 3.1. *Mean weight-for-age and height-for-age* z-*scores of boys and girls at the beginning and end of the study according to socio-economic status*

	Male Mean (*N*)	Female Mean (*N*)	Main effects	Interaction sex and SES
Initial WAZ				
Landless	−2.47 (368)	−2.73 (356)	sex $p = 0.007$	$p = 0.003$
Landowner	−2.49 (301)	−2.46 (313)	SES $p = 0.007$	
Initial HAZ				
Landless	−2.78 (368)	−3.21 (356)	sex $p = 0.005$	$p = 0.002$
Landowner	−2.80 (301)	−2.76 (313)	SES $p = 0.004$	
Final WAZ				
Landless	−2.49 (332)	−2.59 (328)	sex = ns	p = ns
Landowner	−2.43 (277)	−2.42 (298)	SES $p = 0.008$	
Final HAZ				
Landless	−2.71 (332)	−2.89 (328)	sex = ns	p = ns
Landowner	−2.58 (277)	−2.54 (298)	SES $p = 0.001$	
Change in WAZ over 16 months				
Landless	−0.033 (329)	0.12 (321)	sex $p < 0.001$	p = ns
Landowner	0.024 (271)	0.05 (286)	SES p = ns	
Change in HAZ over 16 months				
Landless	0.06 (329)	0.27 (321)	sex $p < 0.001$	p = ns
Landowner	0.17 (271)	0.27 (286)	SES p = ns	

Key: WAZ = weight-for-age *z*-score; HAZ = height-for-age *z*-score; SES = socio-economic status; ns = not significant at 5% level.

and female children according to socio-economic status. The counteracting effect of stunting (low height-for-age) and wasting (low weight-for-age), means that landless female children did not have significantly lower weight-for-height at the beginning of the study compared with other children. Figure 3.3 shows clearly, however, that seasonal weight loss was much greater among landless female children. By September, marking the peak of the monsoon season, landless female children had lost more weight relative to their height than all other children.

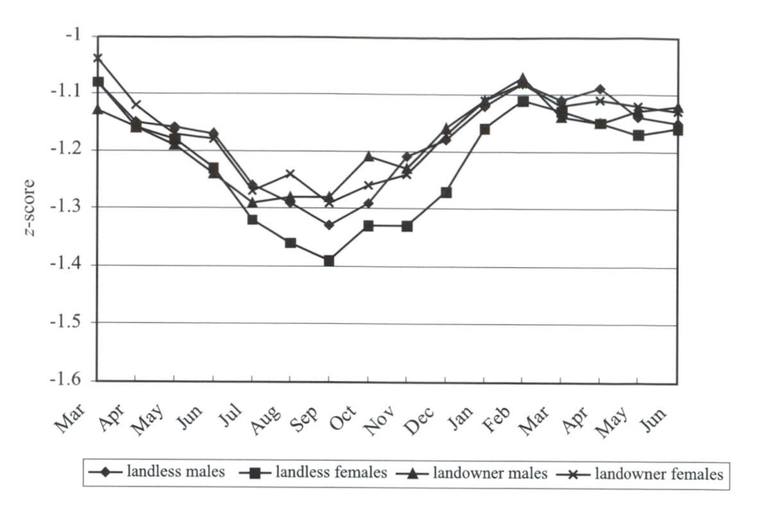

Figure 3.3. Mean weight-for-height of male and female children according to socio-economic status.

Evidence of catch-up growth

Over the 16 months of the study, landless female children grew significantly more than their male counterparts. Table 3.1 shows that the increase in weight-for-age and height-for-age z-scores was significantly greater for landless females than for landless males (effect of sex, $p < 0.001$). This accelerated growth of females took place following the end of the monsoon and a successful rice harvest in late October. Analysis of variance of nutritional status at the end of the study showed that socio-economic status had a significant effect on weight-for-age and height-for-age z scores ($p = 0.008$ and $p = 0.001$ respectively), but sex no longer had a significant effect. Sex differences in nutritional status were, therefore, significant at the beginning of the study but not significant at the end.

Other factors were taken into account in a more detailed analysis of sex differences in growth. After including the effects of family size, birth order, maternal and paternal education, the interaction between sex and socio-economic status remained significant at the beginning of the study and non-significant at the end (Rousham 1996). Contrary to expectations, landless households had fewer children than landowning households (3.6 vs 4.2, $t = -5.75$, $p < 0.001$). Gender bias in landless households at the beginning of the study, therefore, was not an artifact of family size. In terms of child morbidity, food intake and treatment-seeking behaviour, there were no apparent differences between males and females with the exception of diarrhoea which was more commonly reported for females than males. These data relied heavily upon maternal recall, hence it appears that mothers did not under-report symptoms for females, nor preferentially seek treatment for their male children. Children from landless households ate significantly fewer meals per day than those from landowning households, but no sex differences in meal frequency were found (Rousham 1996).

In sum, this study has shown temporal variation in the gender bias in nutritional status. At the beginning of the study, during economic hardship and food shortages, there was a significant gender bias in landless households. At the end of the study, there were no significant gender differences in nutritional status in either socioeconomic group, but there were significant differences between landless and landowning households. Female children from landless households experienced the greatest stunting and wasting during the aftermath of famine and the monsoon season, but displayed catch-up growth towards the end of the study. This indicates that gender discrimination was a response to economic adversity and that discriminatory practices were alleviated once living conditions improved. These findings are supported by previous studies which have shown that landless households are most vulnerable to the effects of natural disasters (D'Souza and Bhuiya 1982) and that female mortality increases disproportionately during severe food shortages (D'Souza and Chen 1980).

Summary

Excess female mortality in South Asia is generally agreed to result from cultural and behavioural practices which give preferential treatment and care to male offspring. The greater value attached to male children can be traced to a complex set of attitudes which are deeply rooted within the predominantly patriarchal family structure and cultural lifestyle of South Asia.

What has emerged from research into the gender bias in South Asia is that female neglect is not universal. Some studies have suggested that there is a conscious discrimination against selected females (Das Gupta 1987, Muhuri and Preston 1991), whereas others have argued that this is only practised through economic necessity (Koenig and D'Souza 1986). It is quite clear, however, that geographical, temporal and socio-economic variations in the treatment of the sexes are extremely important. Factors found to have a significant influence on gender bias in one region, may have no effect in another. These inconsistencies, no doubt, reflect the diversity of culture, behaviour and modes of subsistence within the sub-continent. Efforts to reduce the excess morbidity and mortality of female children must be pursued rigorously but, in order to be successful, these efforts may require different approaches depending on the particular community involved. More research is needed into the sequence of events leading to female death and disadvantage so that sensitive approaches to changing discriminatory attitudes and practices can be developed.

References

Abdullah, M. and Wheeler, E.F. (1985). Seasonal variations, and the intra-household distribution of food in a Bangladeshi village. *American Journal of Clinical Nutrition*, **41**, 1305–13.

Bhuiya, A. and Streatfield, K. (1991). Mothers' education and survival of female children in a rural area of Bangladesh. *Population Studies*, **45**, 253–64.

Chatterjee, M. and Lambert, J. (1989). Women and nutrition: reflections from India and Pakistan. *Food and Nutrition Bulletin*, **11**, 13–28.

Chaudhury, R. H. (1984). Determinants of dietary intake and dietary adequacy for pre-school children in Bangladesh. *Food and Nutrition Bulletin*, **6**, 24–33.

Chen, L. C., Huq, E. and D'Souza, S. (1981). Sex bias in the allocation of food and health care in rural Bangladesh. *Population and Development Review*, **7**, 55–70.

Coale, A. J. (1991). Excess female mortality and the balance of the sexes in the population: an estimate of the number of 'missing females'. *Population and Development Review,* **17**, 517–23.

D'Souza, S. and Bhuiya, A. L. (1982). Socio-economic mortality differentials in a rural area of Bangladesh. *Population and Development Review,* **8**, 753–9.

D'Souza, S. and Chen, L. C. (1980). Sex differentials in mortality in rural Bangladesh. *Population and Development Review,* **6**, 257–70.

Das Gupta, M. (1987). Selective discrimination against female children in rural Punjab, India. *Population and Development Review,* **13**, 77–100.

Dyson, T. and Moore, M. (1983). On kinship structure, female autonomy and demographic behaviour in India. *Population and Development Review,* **9**, 35–60.

Gaur, R. and Singh, N. Y. (1994). Nutritional status among rural Meitei children of Manipur, India. *American Journal of Human Biology,* **6**, 731–40.

Gittelsohn, J. (1991). Opening the box: intrahousehold food allocation in rural Nepal. *Social Science and Medicine,* **33**, 1141–54.

Harriss, B. (1989). Excess female mortality and health care in South Asia. *Journal of Social Studies,* **44**, 1–123.

Harriss, B. (1990). Food distribution, death and disease in South Asia. In *Diet and Disease in Traditional and Developing Societies,* ed. G. A. Harrison and J. C. Waterlow, pp. 290–307. Cambridge: Cambridge University Press.

Hossain, M. M. and Glass, R. I. (1988). Parental son preference in seeking medical care for children less than five years of age in a rural community in Bangladesh. *American Journal of Public Health,* **78**, 1349–50.

Jeyaseelan, L. and Lakshman, M. (1997). Risk factors for malnutrition in South Indian children. *Journal of Biosocial Science,* **29**, 93–100.

Kabeer, N. (1991). Gender dimensions of rural poverty: analysis from Bangladesh. *Journal of Peasant Studies,* **18**, 241–62.

Koenig, M. A. and D'Souza, S. (1986). Sex differences in childhood mortality in rural Bangladesh. *Social Science and Medicine,* **22**, 15–22.

Majumder, A. K., May, M, and Dev Pant, P. (1997). Infant and child mortality determinants in Bangladesh: are they changing? *Journal of Biosocial Science,* **29**, 385–99.

Muhuri, P. K. and Preston, S. H. (1991). Effects of family composition on mortality differentials by sex among children in Bangladesh. *Population and Development Review,* **17**, 415–34.

Okojie, C. (1994). Gender inequalities of health in the Third World. *Social Science and Medicine,* **39**, 1237–47.

Panter-Brick, C. (1997). Seasonal growth patterns in rural Nepali children. *Annals of Human Biology,* **24**, 1–18.

Rahman, M. M., Aziz, K. M. S., Munshi, M. H., Patwari, Y. and Rahman, M. (1982). A diarrhea clinic in rural Bangladesh: influence of distance, age and sex on attendance and diarrheal mortality. *American Journal of Public Health,* **72**, 1124–8.

Rousham, E. K. (1996). Socio-economic influences on gender inequalities in child health in rural Bangladesh. *European Journal of Clinical Nutrition,* **50**, 560–4.

Rousham, E. K. and Mascie-Taylor, C. G. N. (1994). An 18 month study of the effect of periodic anthelminthic treatment on the growth and nutritional status of pre-school children in Bangladesh. *Annals of Human Biology,* **21**, 315–24.

Santow, G. (1995). Social roles and physical health: the case of female disadvantage in poor countries. *Social Science and Medicine,* **40**, 147–61.

Tanner, J. M. (1989). *Foetus into Man.* Ware: Castlemead Publications.

Waldron, I. (1983). Sex differences in human mortality: the role of genetic factors. *Social Science and Medicine,* **17**, 321–33.

Waldron, I. (1987). Patterns and causes of excess female mortality among children in developing countries. *World Health Statistics Quarterly,* **40**, 194–210.

4

Sex, gender and cardiovascular disease

TESSA M. POLLARD

Cardiovascular disease (CVD) is the leading cause of death for women in many industrialised countries. Because more men than women die from CVD (taken here to mean the most important heart diseases, cerebrovascular disease or stroke, and hypertension), however, it is generally identified as a male disease (Brezinka and Padmos 1994, and see Chapter 1). Research in the United States has shown that most women and many doctors are not aware of its importance for women (Reis *et al.* 1997). As a consequence, women were for a long time excluded from clinical trials and epidemiological studies and the results of these investigations were simply extended and applied to women. It is now acknowledged that men and women have differential exposure to risk factors and that male and female bodies respond differently to some of the same risk factors. For instance, the presence of high levels of oestrogen in the bodies of pre-menopausal women is thought to offer them some protection against the development of CVD. It is important to explore these differences rather than make assumptions about women's risks based on data collected from men.

In this chapter, I attempt to demonstrate how gender and sex differences interact to affect the prevalence of CVD in men and women, focusing on the role of oestrogen. I show how modern lifestyles affect levels of oestrogen and how oestrogen may influence women's responses to risk factors for CVD in modern environments. I look in detail at the impact of psychosocial stress on gender differences in CVD and show that stress operates differently for men and women, partly because they are exposed to different stressors (an effect of gender), but also because oestrogen modifies the impact of stress on women's bodies (an effect of sex).

Modernisation and cardiovascular disease

Biological anthropology can contribute to investigations of this type by providing an evolutionary and cross-cultural perspective. Anthropologists' investigations of the human biology of populations living in a variety of different circumstances, from foragers whose lifestyles provide us with the best approximation to that of humans for most of their evolutionary history, to modern city-dwellers, have been used to help us to understand how humans respond to their environments. The assumption, which has now become the basis for the approach labelled Darwinian medicine (Williams and Nesse 1991), is that natural selection has fitted our species to be adapted to the environment in which we spent the vast majority of our evolutionary history, not to the very new circumstances of the modern industrial environment. An understanding of the ways in which we are adapted to the evolutionary environment should help us to see how and why we might be adapted, or maladapted, to other environments. Such an approach is obviously particularly relevant for so called diseases of modernisation, including cancers and CVD.

CVD has long been considered a disease of Westernisation (Trowell and Burkitt 1981). Well-known lifestyle risk factors include a diet high in calories and, particularly, saturated fat and sodium, lack of physical activity and psychosocial stress (Kannel 1987, Pollard 1997). All of these risk factors appear to contribute to the process of atherosclerosis, the build up of fatty and fibrous plaques in the lining of the arteries, which is the main pathology underlying CVDs. All mark departures from life as it must have been for most of our evolutionary history. It seems that our bodies are adapted to cope with far fewer calories than are now routinely available to many, to make maximal use of glucose and fats, to undertake habitual prolonged low to medium level physical activity, and to adapt to relatively rare circumstances requiring a 'fight or flight' response. We are therefore efficient at storing excess calories as fat, and research conducted mainly with men suggests that we respond to the hazards of modern life with what appear to be relatively high levels of stress hormones, such as adrenaline (e.g. James *et al.* 1985). The result is an age-related increase in blood pressure, serum cholesterol levels and atherosclerotic lesions, and thus an increase in the likelihood of the formation of a blood clot in a narrowed artery, blocking the supply of oxygenated blood to part of the brain or heart, causing a stroke or heart attack. Such increases of risk with age are not seen in people living in subsistence societies, whose lifestyles are thought to more closely approximate those of our ancestors.

Modernisation and oestrogen

Within Darwinian medicine little attention has so far been paid to differences between men and women. However, it is mainly biological anthropologists who have explored the regulation of sex hormones in subsistence societies, including foragers. It has become evident that such work is relevant to our understanding of the importance of sex hormones in relation to CVD and certain cancers, since high oestrogen levels increase breast and endometrial cancer risk.

Many of the lifestyle changes that have increased our risk of CVD have also contributed to higher oestrogen levels seen in women in modern industrial environments compared with women living in subsistence-based societies (Ellison *et al.* 1993), partly by causing the ovaries to secrete higher levels of oestrogen during each menstrual cycle, and partly by increasing the number of menstrual cycles that women experience during their lifetimes. Evidence of the effect of the new environment comes from studies of women within modern industrial societies who restrict their dietary intake or engage in regular and rigorous exercise, and also from cross-cultural medical studies and anthropological studies of women who live in societies which are still largely subsistence based.

There is evidence that athletes and women on calorie-restricted diets both show reduced ovarian function compared with other women in modern industrial environments (Ellison *et al.* 1993, Rosetta *et al.* 1998). Medical researchers interested in the role of oestrogen in maintaining population differences in rates of breast cancer have shown that in populations where rates of breast cancer are low, oestrogen levels tend also to be low. They have focused on the role of dietary fats in oestrogen metabolism. Key *et al.* (1990) reported data showing that at pre-menopausal ages, British women had a mean concentration of oestradiol (the most important form of oestrogen) 36% higher than that of Chinese women, and suggest that dietary differences provide the most likely explanation. Women on vegetarian diets, which are usually low in saturated fat, have also been shown to have lower indices of ovarian function than non-vegetarians (Ellison *et al.* 1993).

The development of methods to measure steroid hormones in saliva in the 1980s opened up the possibility for the investigation of sex hormones (progesterone and testosterone, and, with greater difficulty, oestrogens) using samples which can be collected easily in the field. Using these methods in subsistence societies biological anthropologists first demonstrated levels of ovarian hormones that were initially considered to be surprisingly low, and have since shown how levels of these hormones appear to be regulated by the same factors that are known to have an effect within modern industrial environments. Thus, just as physical exercise affects ovarian hormone levels in modern environments, workloads in subsistence agriculturalists have been found to affect ovarian function. For example, among Tamang women in Nepal, Panter-Brick *et al.* (1993) showed that levels of ovarian activity were low compared with levels in US women and that they were most depressed at the time when work in the fields was at its most intensive. Nutritional balance also appears to be important. Studies of weight loss amongst subsistence horticulturalists of the Ituri Forest of Zaire showed that women who lost a lot of weight had lower levels of salivary progesterone and lower frequency of ovulation than women who had lost a smaller amount of weight or had actually gained weight during the same period (Ellison *et al.* 1989).

There is some debate regarding the relative contributions of these factors, but for our purposes what is important is that women whose lifestyles do not allow them to store excess fat tend to have low oestrogen levels. For most of our evolutionary history this would have been an adaptive mechanism preventing repeated pregnancies in women of marginal nutritional status who might have been unable to provide the investment necessary to raise a large number of offspring. These mechanisms operate in the same way in women living in modern environments, despite the fact that for most individuals they can no longer be considered adaptive as a means of fertility regulation.

The wide birth-spacing to which low levels of ovarian function is thought to contribute in subsistence societies further reduces exposure to oestrogen. During most of our species' evolutionary history, the pattern of fertility is thought to have approximated that of present-day foragers, who have a total fertility rate of about 5.6 (Bentley *et al.* 1993). Women would have attained menarche later than is normal in industrial societies today and would have spent much of the time until menopause either pregnant or, mostly, breast-feeding. High levels of oestrogens are secreted by the placenta towards the end of pregnancy, but levels are low during lactational amenorrhoea, which probably lasted up to four years after each birth (Konner and Worthman 1980). In contrast, women in industrial societies today experience menarche at comparatively early ages (Eveleth and Tanner 1990) and give birth to fewer children, often breast-feeding for only a few weeks, or not at all. Women thus experience many more menstrual cycles and associated twice monthly elevations in oestrogen level than previously. The large numbers of women currently taking hormone replacement therapy post-menopausally experience an artificial extension of their cycles and exposure to oestrogen.

One factor characteristic of the modern urban lifestyle and known to increase CVD risk, smoking nicotine, may act to suppress some women's secretion of oestrogen. Smoking is known to be associated with an early menopause (Nilsson *et al.* 1997) and pre-menopausal smokers have lower levels of oestrogen than non-smokers (MacMahon *et al.* 1982). Girdler *et al.* (1997) noted that although smoking prevalence in the United States has declined to around a quarter of the population, the rate of decline in smoking has been much less in women than in men, so a large number of women can be expected to have some reduction in oestrogen levels due to their smoking habits.

In addition to concern about the relationship between high levels of endogenous oestrogen and breast cancer, there have recently been highly publicised worries about the impact of increased levels of exogenous oestrogens in modern industrial environments on the health of both men and women and, particularly, on male fertility (Cadbury 1997, Bentley 1998). Use of oral contraceptives has exposed women to exogenous oestrogen, and there has been some suspicion that women taking oral contraceptives may excrete a form of oestrogen which is still biologically active and which may then enter the water supply. Treatment of several million pregnant women with the powerful synthetic oestrogen diethylstilbestrol (DES) between 1945 and 1971 is another source of exposure and it has been linked to increases in the incidence of undescended testicles and decreased sperm counts in the sons of treated women (Sharpe and Skakkebaek 1993), as well as to rare forms of vaginal cancers and abnormalities of the genitals in their daughters. It has also been suggested that some manufactured compounds, mostly organochlorines, may act as weak oestrogens within the body. Furthermore, some plants contain oestrogens which may exert a biological effect in the human body, but they appear to act to reduce the effects of endogenous oestrogen within the body by competing with it at the receptors which mediate its actions. That is, these phyto-oestrogens may combine with receptors instead of endogenous oestrogens and block their effect. There is currently a great deal of uncertainty regarding the effects of these various exogenous oestrogens within the human body. Golden *et al.* (1998) suggest that in utero exposure to usual levels of environmental oestrogenic substances would be unlikely to be sufficient to cause many of the effects currently attributed to them. Thus, while there is some potential for exposure to exogenous oestrogens within the modern industrial environment to compound, or indeed confound, the effects of endogenous hormone, they will not be considered further.

High levels of endogenous oestrogens do appear to have a number of important implications for women's health. When first identified, ovarian hormones were considered to play a fairly specific role in reproductive physiology. It is now recognised that oestrogen has a number of other important effects throughout the body. As noted earlier, high levels are known to increase the risk of developing

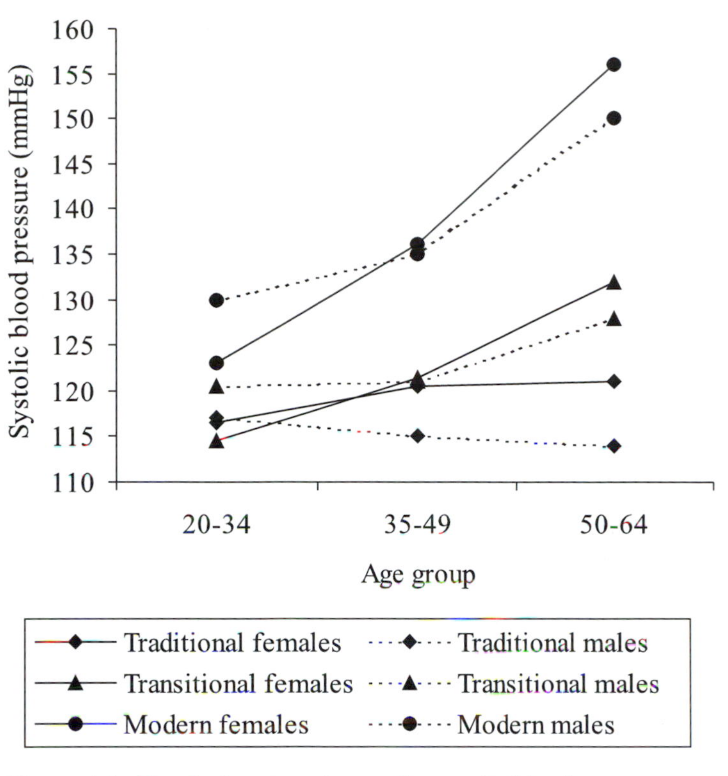

Figure 4.1. Graph showing change in systolic blood pressure with age for men and women from traditional (subsistence-based) societies, transitional or modernising societies, and modern industrial societies. Modified from Pollard *et al.* (1991).

breast cancer, and, post-menopausally, low levels of oestrogen are implicated in osteoporosis and Alzheimer's disease (Legato 1997). Fluctuations in oestrogen levels have been linked to mood changes in women, both pre-menstrually and during the menopause, although some have questioned the evidence that women's moods are affected in this way (Nicolson 1995).

Evidence that oestrogen may have a role in protecting against CVD comes from a variety of sources. Women's risk of CVD increases after menopause and there is now good evidence that the reason for this is the post-menopausal drop in oestrogen levels (Knopp *et al.* 1994, Haffner and Valdez 1995). One reason for the higher risk is the increase in blood pressure of women at and after menopause. This increase is seen in all types of society, as shown in Figure 4.1, but only becomes clinically important in modernising and modern societies, where blood pressure levels are much higher than in subsistence societies. Studies of post-menopausal women who maintain higher levels of oestrogen through hormone replacement therapy show that they have a reduced risk of developing CVD compared with unmedicated women (Reis *et al.* 1997).

Researchers have identified a number of physiological mechanisms by which oestrogen may have this effect. For example, oestrogen appears to affect serum cholesterol levels favourably, and to retard the development of atherosclerosis in a number of ways (Karas *et al.* 1994, Chester *et al.* 1995). It has been suggested that oestrogen inhibits the oxidation of low-density lipoproteins (a process which contributes to atherosclerosis) and acts as a vasodilator, preventing the constriction of arteries (Reis *et al.* 1997).

We should not forget that an increase in ovarian function is also reflected in higher levels of progesterone (Ellison *et al.* 1993). Progestins (of which progesterone is one), given as part of hormone replacement therapy to reduce the risk of endometrial cancer associated with oestrogen supplementation, appear to reduce the cardio-protective effects of oestrogen to some extent. For example, they act to block the beneficial effect of oestrogen on serum lipoproteins (Reis *et al.* 1997). Progestins also reduce the vasodilator effect of oestrogen. It is important, therefore, to remember that oestrogen does not act in isolation and that its effects may be counteracted to some extent by high levels of progestins. However, relatively little research has been conducted on the effects of progestins and for the moment the focus remains on oestrogen.

To summarise, there is good evidence that many risk factors for CVD also contribute to a higher secretion of oestrogen in women, which seems to offer some protection to women that is not available to men. This protection may not be so strong for women living in developing societies who are exposed to CVD risk factors in adulthood that were not present in their childhood environments, as happens during urbanisation. Ellison (1996) has proposed that differences in adult gonadal function may be at least partly due to developmental processes operating during childhood and adolescence to establish adult set-points of the hormonal axis involved. If this is the case, women newly exposed to a modern environment during their lifespan may not be as well protected as women living in a more stable industrial environment (Pollard 1997).

Gender, stress and the role of oestrogen

Stress is now generally acknowledged as a risk factor for CVD (James *et al.* 1987, Krantz *et al.* 1988). In the remainder of this chapter I shall review research which throws light on differences between the ways in which stress may act to alter CVD risk in men and women. Gender differences in experiences of stress have recently been given some detailed consideration in the literature, correcting a previous bias towards the study of men. Recently, also, the unique role that oestrogen may play in moderating responses to stress has been given some attention. Here is another instance where oestrogen may provide women with protection against the modern lifestyle which is not available to men. I shall examine in some detail the importance of gender and sex differences with respect to links between stress and CVD, and the interactions between them, because they provide a useful illustration of the important ways in which social and biological factors act together to influence disease risk. Thus in the following sections I consider whether men and women are likely to have different experiences of stressors, and differences in their responses to stress.

Gender differences in experiences of stressors

The most common definition of stress is as a perception of excessive demands with which an individual is unable to cope (Lazarus 1966). Given the powerful role of gender in determining social roles, it is not surprising that men and women are often exposed to different stress-generating experiences (stressors), and that they often show different emotional responses to them. In other words, women and men are very likely to encounter different stressors in their day to day lives, and there is some evidence that their perceptions of the stressfulness of these experiences may also differ. These differences have been explored by psychologists and sociologists interested in explaining the consistent reporting by women of higher levels of distress than those reported by men, as demonstrated by a variety of measures (see Littlewood, Chapter 8).

It is notable that much of the earliest research linking stress and CVD focused on the effects of the demands experienced by men in management positions. In particular, Type A behaviour, a label applied to people who are competitive, hard driving, hurried and aggressive, and linked to an increased risk of CVD, was typically identified with male managers with high levels of responsibility and a heavy workload. More recently the effects of lack of control have been the focus of studies on stress, and there is good evidence to suggest that women experience less control than men, particularly at work (Hall 1989). Women generally occupy lower status jobs, offering them less control, less security and less financial reward than men receive. There has also been a shift towards the consideration of stressors outside the workplace, and an acknowledgement that women often take on a greater burden in the household than do men, even when they are in paid employment. Thus most women are likely to encounter more stressors than most men during daily life.

There is an argument that women report greater stress than men because they express their emotional distress more than men. It is obviously extremely difficult to disentangle gender effects on day to day experiences and on the reporting of these experiences. However, Mirowsky and Ross (1995) attempted an analysis which was designed to test whether the greater distress of women was a consequence of differences in men and women's reporting styles, and concluded that although there are such gender differences, they cannot on their own explain the greater distress experienced by women. In a study which did not focus on stress but on complaints about common cold symptoms, which were compared against ratings of those symptoms by a medical observer, Macintyre (1993) found that men rated their symptoms to be more severe than did women. Thus it appears that women are not necessarily socialised to complain more than men, although it may be that men and women feel differently about the acceptability of complaining about emotional problems versus physical symptoms.

It is undoubtedly true that gender affects experiences of stress in several ways. It is probably fair, however, to suggest that evidence is accumulating that women report more stress, on average, than men because they are exposed to more stressful experiences, at least in modern industrialised societies where most of the relevant research has been conducted. We might therefore expect stress to be a more important risk factor for CVD in women than in men. However, the opposite appears to be true when we consider the effects of biological differences.

Do men and women show different physiological responses to stress?

If a man or woman feels stressed, what are the physiological consequences and are there sex differences in these consequences with implications for cardiovascular health? That is, what is the biological effect of the perception of stress for men and women?

Oestrogen may have a role to play here, influencing the sympathetic nervous system and the effects of the main so-called 'stress hormones', adrenaline and cortisol. The secretion of adrenaline, a catecholamine, from the adrenal medulla is controlled by the sympathetic nervous system. It stimulates the release of metabolites that can be used to provide immediate sources of energy, such as glucose and free fatty acids, in the once adaptive 'fight or flight' mechanism, mentioned above. Free fatty acids may contribute to elevated serum cholesterol levels if not utilised for energetic requirements, as they would be if an individual were to fight or flee. In addition, adrenaline causes platelets to become more cohesive, increasing the risk of clot formation and thus a heart attack or stroke. Cortisol is secreted by the adrenal cortex and its levels are controlled by other hormones. It also acts to increase the breakdown of fats to free fatty acids and, in addition, stimulates the formation of adrenaline and inhibits its breakdown (Toates 1995). However, while cortisol rises in response to stress in laboratory situations, it is not clear how its regulation is influenced by 'real-life' stress experienced on a daily basis (Pollard 1995).

Both hormones, but particularly adrenaline, are involved in the regulation of blood pressure, which is also affected by more direct actions of the autonomic nervous system. It is thought that repeated elevation of blood pressure may lead to sustained hypertension, possibly as a result of structural changes in the blood vessels and damage to the arterial walls, which may precipitate atherosclerosis. Blood pressure may be raised by an increase in either cardiac output (the quantity of blood pumped through the heart in a given time, determined by heart rate and stroke volume), or by an increase in the resistance of the arterial system, known as total peripheral resistance. Polefrone and Manuck (1987) noted that circulating adrenaline has a less important effect on cardiac performance than do sympathetic neural influences, so that cardiac changes can occur independently from changes in adrenaline levels.

If stress responses are, as seems very likely, related to the development of CVD, any differences between men and women in this respect may help us to understand the gender difference in rates of these diseases. As in other areas, most early studies in which physiological responses to stress in humans have been assessed were conducted using only men as subjects. When researchers began to include women in such studies they found that women's responses were not always the same as men's (Stoney *et al.* 1987). Frankenhaeuser, who made most of the initial findings, has suggested that the differences were related to gender roles, that is, to different emotional responses in the face of the challenges being studied (Frankenhaeuser 1991). She considered that young Swedish women showed a smaller physiological response to exams than men in the 1970s because they were not as achievement-oriented as young men at that time. She also reported findings that female students in male dominated courses or in male dominated occupations had more 'male' endocrine responses to stress. Her group found that mothers and fathers who took their three year old children to hospital for a medical check up secreted similar amounts of adrenaline and cortisol, and suggested this is because women are expected to show more competence in this sphere, though it is noteworthy that their hormone levels were not *higher* than men's, even in this situation which is thought to evoke greater stress in women (Lundberg *et al.* 1981). Thus Frankenhaeuser has argued that it is affective responses which differ between men and women, as discussed above, rather than physiological responses. Obviously it is important to investigate both types of response in such studies in order to resolve this question.

Later work appears to suggest that there are real physiological differences between men and women. Thus some health psychologists have started to investigate the possibility that men and women may show different physiological responses to similar feelings of stress. Polefrone and Manuck (1987) reviewed findings available at the time. They noted that with respect to neuroendocrine responses to stress there appeared to be fairly consistent evidence from laboratory studies that men show a greater adrenaline response than women. In a meta-analysis published at the same time, Stoney *et al.* (1987) came to a similar conclusion. The findings with respect to cortisol are much less clear-cut and will not be considered further since cortisol's role as a stress hormone is anyway less well understood and less obviously related to cardiovascular health.

There is some evidence that post-menopausal women lose this apparent protection against elevation of adrenaline (Saab *et al.* 1989) due to low levels of oestrogen. Furthermore, women given oestrogen either orally or through skin patches (as in hormone replacement therapy) have been found to show blunted adrenaline responses to stressors, compared with control subjects who have not received exogenous oestrogen (Lindheim *et al.* 1992, del Rio *et al.* 1994). There has been less success in identifying any change in neuroendocrine reactivity at different phases of the menstrual cycle (Stoney *et al.* 1990, Litschauer *et al.* 1998), perhaps because the differences in hormone levels are not as marked. Nevertheless, it seems certain that men show a stronger 'fight or flight' response to stress than women in laboratory situations, and it is very likely that oestrogen is responsible for the comparative blunting of women's responses. Or, put differently, the fact that men do not have high levels of circulating oestrogen means that they show a greater adrenaline response to stress.

The findings described above stem from research undertaken in laboratory conditions. As Matthews *et al.* (1991) note, 'to be a useful construct for understanding sex differences in coronary heart disease, sex differences in stress responses should be apparent during naturally occurring stressors, as well as during laboratory stressors'. We set out to test for such a difference between men and women during their normal working lives. In a study involving both men and women in a variety of mostly non-manual occupations, we found that adrenaline levels were significantly higher in men on days when they were at work than at the weekend, but observed no such difference in women (Figure 4.2). Furthermore, statistical analysis indicated that variation in self-reported demand experienced on rest and work days was correlated with the adrenaline response seen in men, but there was no such correlation in women (Pollard *et al.* 1996). The difference in adrenaline response was seen despite the fact that men and women reported (by questionnaire) very similar levels of demand at the weekend and on the work days; both reported much higher demand on work days (Figure 4.2). Importantly, they also reported similar mood states, with the exception that men's anxiety dropped just after work, whereas women's did not (Figure 4.2). While both men and women reported an increase in workload on working days, and associated increases in anxiety, only men showed a corresponding increase in adrenaline levels. We interpreted these results to suggest that differences in relevant subjective experience could not explain the difference in adrenaline response to work between men and women, thus implicating biological mechanisms.

With respect to the responses of the cardiovascular system itself, Stoney *et al.*'s (1987) meta-analysis also identified a sex difference

Figure 4.2. Graphs showing changes from Sunday (a rest day) to Tuesday (a work day) in urinary adrenaline excretion rate and stress-related perceptions over the same days for working women (N = 53, solid lines) and men (N = 51, dotted lines). The change in urinary adrenaline excretion rate (measured at 6 pm, reflecting hormone secretion during the afternoon) over days was significant in men but not women. There were no statistically significant differences between men and women in perceived demand or mood at any time (Pollard *et al.* 1996).

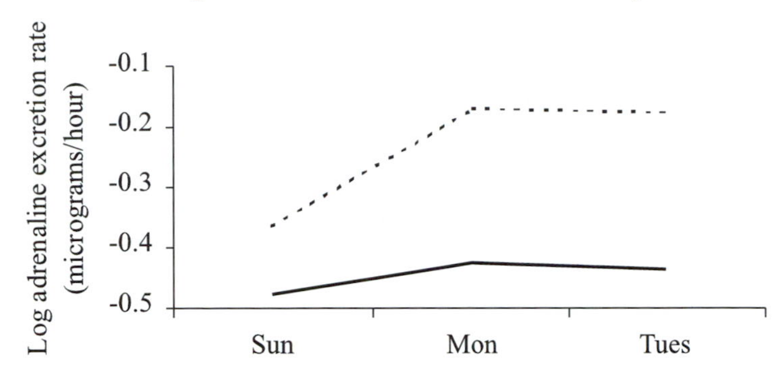
Change in adrenaline excretion over 3 days
Log adrenaline excretion rate
(micrograms/hour)
-0.1
-0.2
-0.3
-0.4
-0.5
Sun
Mon
Tues

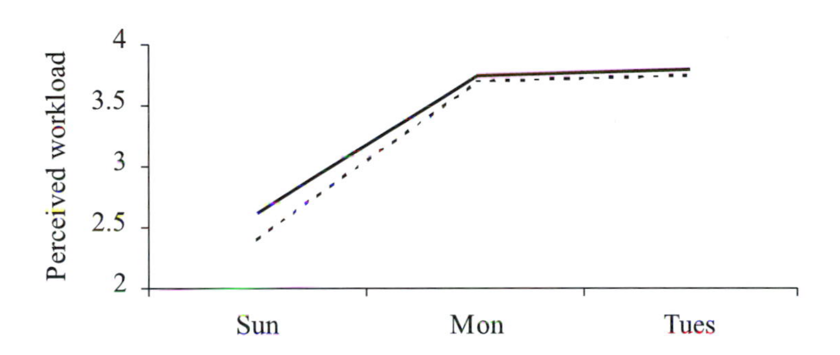
Change in perceived workload over 3 days
Perceived workload
4
3.5
3
2.5
2
Sun
Mon
Tues

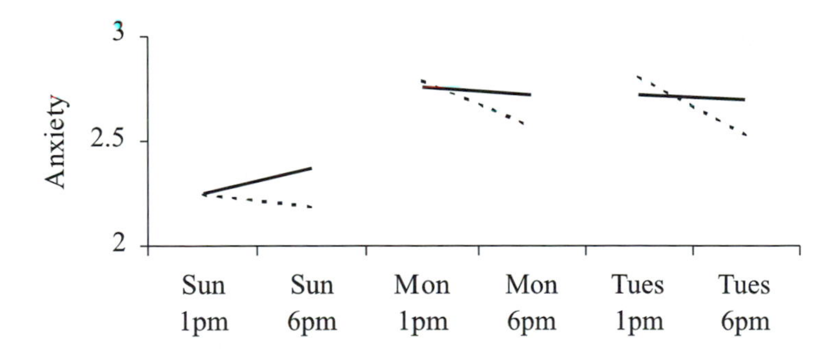
Change in anxiety over 3 days
Anxiety
3
2.5
2
Sun 1pm
Sun 6pm
Mon 1pm
Mon 6pm
Tues 1pm
Tues 6pm

in systolic blood pressure reactions to stressors, with men showing a bigger response than women. Furthermore, post-menopausal women show greater systolic blood pressure responses than pre-menopausal women (Saab *et al.* 1989, Merz *et al.* 1998). In contrast, women with high levels of circulating oestrogen appear to show a bigger increase in heart rate in response to stress. More recently, studies have looked in greater detail at the components of the cardiovascular response and found evidence that pre-menopausal women tend to show a greater myocardial response, as indexed by heart rate and cardiac output (partly determined by heart rate), whereas men tend to show a greater vascular response, with greater increases in total peripheral resistance (Girdler *et al.* 1997). It is thought that the latter may be more dangerous with respect to CVD. Girdler and Light (1994) suggested that the greater reactivity of the heart seen in women may not increase the risk of CVD if the blood vessels supplying the heart are relaxed and dilated. Again, this difference in response may be influenced by the actions of oestrogen, as evidenced by the fact that women with higher levels of oestrogen appear to show greater stroke volume or cardiac output and lesser total peripheral resistance during stress (Girdler *et al.* 1997). Girdler *et al.* (1997) found that women smokers show a lesser increase in cardiac output and heart rate and a greater increase in total peripheral resistance in response to laboratory stressors than do women non-smokers, that is, a more 'male', and more risky, pattern of response. They suggest that suppression of oestrogen caused by smoking is the cause. Oestrogen may enhance the contractivity of the heart (Girdler and Light 1994), which would increase stroke volume and thus cardiac output.

There is also some evidence that men show greater short-term elevation in serum lipid levels in response to stress (Matthews *et al.* 1991, Niaura *et al.* 1992). In general, then, there are now compelling data available to suggest that men and women show different physiological responses to stress, and to implicate oestrogen as an important explanation of these differences. It is worth noting, however, that there has recently been some suggestion that progestins may counteract effects of oestrogen with respect to physiological responses to stress (Lindheim *et al.* 1994), such that the generally positive effect of oestrogen may be reduced when progestin levels are high. It has also been suggested that the balance of androgens and oestrogens may be important for both men and women (Litschauer *et al.* 1998). Nevertheless, women do appear to have some advantage over men in that the way in which their bodies respond to stress does not increase their risk of developing CVD as much as it does for men. Matthews *et al.* (1991) have also pointed out that even when women do show immediate physiological responses to stress, the actions of oestrogen may protect them from the long-term negative effects suffered by men.

The impact of stress on oestrogen levels

So far we have only considered the potential impact of stress on cardiovascular health and the way in which oestrogen may modify that impact. However, stress may also affect oestrogen levels, both directly and indirectly. Scientists interested in explaining the success rate of couples who are trying to conceive have suggested that psychosocial stress may affect ovarian function and there is some evidence that women are more likely to conceive during months in which they report less stress (Saunders and Bruce 1997). It may be that stress suppresses hypothalamic gonadotropin-releasing hormone (GnRH), which has an important role in regulating the pituitary–ovarian axis (Berga 1995). Evidence to support the existence of such a mechanism comes from studies of non-human primates such as *Macaca fascicularis*, in which social subordinates are considered to experience more stress and also show ovarian dysfunction (Shively *et al.* 1997).

Stress may also have an indirect impact on oestrogen levels through changes in smoking behaviour. Women are particularly likely to say that they smoke in response to distress, and that when they stop smoking stressful events trigger a relapse (Biener 1987). It has been suggested that nicotine may affect mood more strongly for women than for men because of sex differences in how it is metabolised (Biener 1987), so the roots of this difference may possibly be partly biological. As noted above, smoking appears to reduce oestrogen levels.

Conclusion

As cultural innovations have moved humans further away from the physical and social environments in which they spent most of their evolutionary history, they have been exposed to a variety of new threats to their health, some of which have arisen because of the maladaptation of the human body to new (in evolutionary terms) environments. Research by biological anthropologists has demonstrated how ovarian function responds to various demands made on women's bodies. It seems that natural selection has created a mechanism whereby women whose nutritional status is challenged either by excessive energy expenditure or by insufficient energy intake reduce their secretion of sex hormones to make conception, and thus the birth of an offspring that they may not be able to care for, less likely. Conversely, women in modern industrial environments, whose energy expenditure is low and whose energy intake is often greater than their needs, secrete much higher levels of oestrogen and they may also expose themselves to exogenous oestrogens by taking oral contraceptives or hormone replacement therapy. Exposure of women's bodies to high levels of oestrogens is a new phenomenon in evolutionary terms, and we are only just beginning to understand its consequences for health.

The increase in circulating oestrogen, resulting partly from the same risk factors that increase CVD risk, provide women in modern industrial environments with some degree of protection which is not available to men, partly explaining the lower rates of CVD in women. The interaction between lifestyle change, gender roles and sex differences in biology is complex, as is well illustrated when we consider the effects of stress on CVD for men and women. First, men and women are likely to be exposed to different stressors, and they are socialised to respond differently, both emotionally and behaviourally. Second, oestrogen probably blunts risky physiological responses to stress in women. So although women's lives seem often to be more stressful than men's, they are protected by high levels of oestrogen associated with modern industrial lifestyles. However, stress may also reduce oestrogen levels, so the interactions between stress and oestrogen cannot be considered simple. We should note here that women who reduce their oestrogen levels through healthy lifestyle choices, that is a low fat diet and exercise rather than smoking and exposure to stressors, will benefit from a simultaneous reduction in CVD risk and the risk of developing cancers associated with high oestrogen levels.

To understand the role played by stress as a risk factor for CVD we have to conduct studies of both men and women. Research results obtained from studies on men only cannot simply be extended to explain the effects of stress on women's bodies. Moreover, future research on this topic must also take into account the importance of gender as manifested in social contexts as well as the biological differences between men and women, and must consider the impact of the interaction between the two.

References

Bentley, G. R. (1998). Environmental pollutants and infertility. *Journal of Biosocial Science,* **30**, 277.

Bentley, G. R., Jasienska, G. and Goldberg, T. (1993). Is the fertility of agriculturalists higher than that of nonagriculturalists? *Current Anthropology,* **34**, 778–85.

Berga, S. L. (1995). Stress and amenorrhea. *Endocrinologist,* **5**, 416–21.

Biener, L. (1987). Gender differences in the use of substances for coping. In *Gender and Stress,* ed. R. Barnett, L. Biener and G. Baruch, pp. 330–49. New York: Free Press.

Brezinka, V. and Padmos, I. (1994). Coronary heart disease risk factors in women. *European Heart Journal,* **15**, 1571–84.

Cadbury, D. (1997). *The Feminization of Nature: Our Future at Risk.* London: Penguin.

Chester, A. H., Jiang, C., Borland, J. A., Yacoub, M. H. and Collins, P. (1995). Oestrogen relaxes human epicardial coronary arteries through non-endothelium-dependent mechanisms. *Coronary Artery Disease,* **6**, 417–22.

del Rio, G., Velardo, A., Zizzo, G. *et al.* (1994). Effect of estradiol on the sympathoadrenal response to mental stress in normal men. *Journal of Clinical Endocrinology and Metabolism,* **79**, 836–40.

Ellison, P. T. (1996). Age and developmental effects on human ovarian function. In *Variability in Human Fertility*, ed. L. Rosetta and C.G.N. Mascie-Taylor, pp. 69–90. Cambridge: Cambridge University Press.

Ellison, P. T., Peacock, N. R. and Lager, C. (1989). Ecology and ovarian function among Lese women of the Ituri Forest, Zaire. *American Journal of Physical Anthropology*, **78**, 519–26.

Ellison, P. T., Panter-Brick, C., Lipson, S. F. and O'Rourke, M. T. (1993). The ecological context of human ovarian function. *Human Reproduction*, **8**, 2248–58.

Eveleth, P. B. and Tanner, J. M. (1990). *Worldwide Variation in Human Growth*. Cambridge: Cambridge University Press.

Frankenhaeuser, M. (1991). The psychophysiology of sex differences as related to occupational status. In *Women, Work and Health*, ed. M. Frankenhaeuser, U. Lundberg and M. Chesney, pp. 39–61. New York: Plenum.

Girdler, S. S. and Light, K. C. (1994). Hemodynamic stress responses in men and women examined as a function of female menstrual cycle phase. *International Journal of Psychophysiology*, **17**, 233–48.

Girdler, S. S., Jamner, L. D., Jarvik, M., Soles, J. R. and Shapiro, D. (1997). Smoking status and nicotine administration differentially modify hemodynamic stress reactivity in men and women. *Psychosomatic Medicine*, **59**, 294–306.

Golden, R. J., Noller, K. L., Ernstoff, L. T. *et al.* (1998). Environmental endocrine modulators and human health: an assessment of the biological evidence. *Critical Reviews in Toxicology*, **28**, 109–227.

Haffner, S. M. and Valdez, R. A. (1995). Endogenous sex hormones: impact on lipids, lipoproteins, and insulin. *American Journal of Medicine*, **98 (suppl 1A)**, S40–7.

Hall, E. M. (1989). Gender, work control, and stress: a theoretical discussion and an empirical test. *International Journal of Health Services*, **19**, 725–45.

James, G. D., Jenner, D. A., Harrison, G. A. and Baker, P. T. (1985). Differences in catecholamine excretion rates, blood pressure and lifestyle among young Western Samoan men. *Human Biology*, **57**, 635–47.

James, S. A. (1987). Psychosocial precursors of hypertension: a review of the epidemiologic evidence. *Circulation*, **76 (suppl 1)**, I60-6.

Kannel, W. B. (1987). New perspectives on cardiovascular risk factors. *American Heart Journal*, **114**, 213–19.

Karas, R. H., Patterson, B. L. and Mendelsohn, M. E. (1994). Human vascular smooth muscle cells contain functional estrogen receptor. *Circulation*, **89**, 1943–50.

Key, T. J. A., Chen, J., Wang, D. Y., Pike, M. C. and Boreham, J. (1990). Sex hormones in women in rural China and in Britain. *British Journal of Cancer*, **62**, 631–6.

Knopp, R. H., Zhu, X. and Bonet, B. (1994). Effects of estrogens on lipoprotein metabolism and cardiovascular disease in women. *Atherosclerosis*, **110 (suppl)**, S83-91.

Konner, M. and Worthman, C. (1980). Nursing frequency, gonadal function, and birth spacing among the !Kung hunter–gatherers. *Science*, **207**, 788–91.

Krantz, D. S., Contrada, R. J., Hill, R. and Friedler, E. (1988). Environmental stress and biobehavioral antecedents of coronary heart disease. *Journal of Consulting and Clinical Psychology,* **56**, 333–41.

Lazarus, R. S. (1966). *Psychological Stress and the Coping Process.* New York: McGraw-Hill.

Legato, M. J. (1997). Gender-specific physiology: How real is it? How important is it? *International Journal of Fertility,* **42**, 19–29.

Lindheim, S. R., Legro, R. S., Bernstein, L. *et al.* (1992). Behavioral stress responses in premenopausal and postmenopausal women and the effects of estrogen. *American Journal of Obstetrics and Gynecology,* **167**, 1831–6.

Litschauer, B., Zauchner, S., Huemer, K. H. and Kafka-Lutzow, A. (1998). Cardiovascular, endocrine, and receptor measures as related to sex and the menstrual cycle phase. *Psychosomatic Medicine,* **60**, 219–26.

Lundberg, U., de Chateau, P., Winberg, J. and Frankenhaeuser, M. (1981). Catecholamine and cortisol excretion patterns in three-year-old children and their parents. *Journal of Human Stress,* **7**, 3–11.

Macintyre, S. (1993). Gender differences in the perceptions of common cold symptoms. *Social Science and Medicine,* **36**, 15–20.

MacMahon, B., Trichopoulos, D., Cole, P. and Brown, J. (1982). Cigarette smoking and urinary estrogens. *New England Journal of Medicine,* **15**, 1062–5.

Matthews, K. A., Davis, M. C., Stoney, C. M., Owens, J. F. and Caggiula, A. R. (1991). Does the gender relevance of the stressor influence sex differences in psychophysiological responses? *Health Psychology,* **10**, 112–20.

Merz, C. N. B., Kop, W., Krantz, D. S., Helmers, K. F., Berman, D. S. and Rozanski, A. (1998). Cardiovascular stress response and coronary artery disease: evidence of an adverse postmenopausal effect in women. *American Heart Journal,* **135**, 881–7.

Mirowsky, J. and Ross, C. E. (1995). Sex differences in distress: real or artifact. *American Sociological Review,* **60**, 449–68.

Niaura, R., Stoney, C. M. and Herbert, P. N. (1992). Lipids in psychological research: the last decade. *Biological Psychology,* **34**, 1–43.

Nicolson, P. (1995). The menstrual cycle, science and femininity: assumptions underlying menstrual cycle research. *Social Science and Medicine,* **41**, 779–84.

Nilsson, P., Moller, L., Koster, A. and Hollnagel, H. (1997). Social and biological predictors of early menopause: a model for premature aging. *Journal of Internal Medicine,* **242**, 299–305.

Panter-Brick, C., Lotstein, D. S. and Ellison, P. T. (1993). Seasonality of reproductive function and weight loss in rural Nepali women. *Human Reproduction,* **8**, 684–90.

Polefrone, J. M. and Manuck, S. B. (1987). Gender differences in cardiovascular and neuroendocrine response to stressors. In *Gender and Stress,* ed. R. Barnett, L. Biener and G. Baruch, pp. 13–38. New York: Free Press.

Pollard, T. M. (1995). Use of cortisol as a stress marker: practical and theoretical problems. *American Journal of Human Biology*, **7**, 265–74.

Pollard, T. M. (1997). Environmental change and cardiovascular disease: a new complexity. *Yearbook of Physical Anthropology*, **40**, 1–24.

Pollard, T. M., Brush, G. and Harrison, G. A. (1991). Geographic distribution of within-population variability in blood pressure. *Human Biology*, **63**, 643–61.

Pollard, T. M., Ungpakorn, G., Harrison, G. A. and Parkes, K. R. (1996). Epinephrine and cortisol responses to work: a test of the models of Frankenhaeuser and Karasek. *Annals of Behavioral Medicine*, **18**, 229–37.

Reis, S. E., Holubkov, R. and Zell, K. A. (1997). Women's hearts are different. *Current Problems in Obstetrics, Gynecology and Fertility*, **20**, 73–92.

Rosetta, L., Harrison, G. A. and Read, G. F. (1998). Ovarian impairments of female recreational distance runners during a season of training. *Annals of Human Biology*, **25**, 345–58.

Saab, P. G., Matthews, K. A., Stoney, C. M. and McDonald, R. H. (1989). Premenopausal and postmenopausal women differ in their cardiovascular and neuroendocrine responses to behavioral stressors. *Psychophysiology*, **26**, 270–80.

Saunders, K. A. and Bruce, N. W. (1997). A prospective study of psychosocial stress and fertility in women. *Human Reproduction*, **12**, 2324–9.

Sharpe, R. M. and Skakkebaek, N. E. (1993). Are oestrogens involved in falling sperm counts and disorders of the male reproductive tract? *Lancet*, **341**, 1392–5.

Shively, C. A., Laber-Laird, K. and Anton, R. F. (1997). Behavior and physiology of social stress and depression in female cynomolgus monkeys. *Biological Psychiatry*, **41**, 871–82.

Stoney, C. M., Davis, M. and Matthews, K. A. (1987). Sex differences in physiological responses to stress and in coronary heart disease. *Psychophysiology*, **24**, 127–31.

Stoney, C. M., Owens, J. F., Matthews, K. A., Davis, M. C. and Caggiula, A. (1990). Influences of the normal menstrual cycle on physiologic functioning during behavioral stress. *Psychophysiology*, **27**, 125–35.

Toates, F. (1995). *Stress: Conceptual and Biological Aspects*. Chichester: Wiley.

Trowell, H. C. and Burkitt, D. P. (ed.) (1981). *Western Diseases: Their Emergence and Prevention*. Cambridge, Mass.: Harvard University Press.

Williams, G. C. and Nesse, R. M. (1991). The dawn of Darwinian medicine. *Quarterly Review of Biology*, **66**, 1–22.

5

Social meanings and sexual bodies: gender, sexuality and barriers to women's health care

LENORE MANDERSON

In this chapter I explore the impact of cultural constructions of gender and sexuality on women's and men's health. Drawing on work conducted in subsistence, industrialising and industrialised societies, particularly with women, I illustrate the ways in which sexual, reproductive and other health problems are influenced by ideas about and relations of gender and sexuality. Despite differences in social, cultural and economic contexts, women everywhere are subordinate to men to some degree. Their sexuality is controlled and their social value is tied to their reproductive abilities, both in terms of having children and maintaining the household. These factors influence women's abilities to care for their bodies and to control their fertility. The sexual meanings ascribed to women's bodies influence popular interpretations of the signs and symptoms of health and illness and affect help-seeking behaviour and access to health services and treatment, especially for ailments which might be construed as sexual. Gender also affects men's access to and use of health services in relation to sexual and urinary tract infections in many communities and for that reason, I include some discussion of men's as well as women's health in this chapter.

The social context of health and illness

While this chapter focuses on the social experiences of illness, care and outcome, at the outset it is important to note that the risk of infection and disease occurs through a combination of biological, social and structural factors. Changes to local ecologies by development programmes, patterns of population movement and settlement, water and sanitation, all affect the prevalence of vectors and microbes and, hence, influence the potential distribution of infectious disease for both women and men. Urbanisation and industrialisation further affect health through exposure to different pathogens in the environment and as a result of conditions of labour (e.g. Pick and Obermeyer 1996, Khanna 1997), which are, in turn, mediated by social variables including class, age and gender. Gender, for example, determines individuals' interactions with, movement through and use of the environment, their involvement in productive and reproductive tasks, and their use of medical technologies (Khanna 1997). Women are usually at greater risk of infection from schistosomiasis (bilharzia) than men because they must often use snail-infested water sources for extended periods in order to wash clothes, kitchen utensils, themselves and their children. Men's exposure, on the other hand, is more limited (Huang and Manderson 1992).

Gender and kinship relations affect women's mobility, influencing the kinds of illness they are most likely to report and the opportunities they have to receive medical care (Puentes-Markides 1992). Despite some changes with development and industrialisation, gender everywhere remains a major factor accounting for 'significant disparities' in the health of women and men (Ojanuga and Gilbert 1992:613). Women's lower levels of education and limited access to information influence their ability to recognise the signs of illness, to impute those signs with meaning, and to act consequentially. Gender relations structure household responsibilities and priorities and access to finances and other assets which may determine treatment or predispose women's willingness to expend resources on their own health. Ideologies of gender, motherhood and family all determine the care provided to women for their own health, to their children and to other household members. Power relations within households also influence the distribution of resources, as the following example from South Asia suggests.

> Gita was seven months pregnant. She had very swollen ankles and felt dizzy and ill, and asked her mother-in-law for permission to attend the local health centre. Her mother-in-law refused; she said that Gita was lazy and was using her pregnancy as an excuse to do

> even less housework than usual. Her son supported his mother; she knew more than he about women's illness. The intervention of a local community worker on Gita's behalf only made him angry. Gita's constant weeping and complaining irritated her affines, however, and eventually she was given enough money to travel three days by bus to her natal village. There, she could receive medical attention and personal care from her own mother, until after the birth of the baby.
>
> (Pike 1997, personal communication)

Reproductive imperatives

Men's status is culturally determined through their roles in production and in public life. Marking their transition to adulthood, however, may be sexual. For example, initiation practices may involve sexual relations among men (Herdt 1982) or circumcision and subincision rites which physically inscribe adulthood on men's bodies (Willis 1997); and, in some societies, young men's sexual activity is 'initiated' commercially, as in Thailand where visits to prostitutes are routine (Sukanya 1988, cf. Lyttleton 1994). Sexual expression is a component of manhood, as reflected by various anxieties regarding semen loss (Bottero 1991), proscriptions on frequency of intercourse, anxieties regarding sexual function and fear of loss of virility manifested through such health problems as *koro*, which is the dramatic retraction of the male sexual organs (Yap 1977, Edwards 1985). There is little documentation of cultural disapprobation of male infertility, however, as it is almost always the woman who is assumed to be infertile in a barren marriage.

Men's sexuality is often acknowledged through sexual initiation, and their social status derives from cultural achievement; in contrast, women's status derives primarily from their reproductive role, either by bearing large numbers of children or, in patrilineal societies especially, by producing sons. The importance of reproduction in determining women's status applies across cultures and geographic regions, and the relationship of childbearing to women's subordinate status was an early concern of feminist anthropology (see contributions in Rosaldo and Lamphere 1974, Handwerker 1990, Greenhalgh 1995, Rice and Manderson 1996). However, childbearing also grants a woman adult status and some authority within her household, kin group and community (e.g. Ram 1991, Defo 1997). Where women's material well-being and social status depends upon her demonstrated reproductive capacity, 'barren' women may be divorced or forced to accept a polgynous union (Bledsloe 1995, Jennaway 1996, Symonds 1996).

The social value of motherhood limits the acceptance of contraception. An example from rural West Africa highlights the pressure on women to reproduce and the costs to those who try to limit family size.

> Lardi was 33 years old and had six children. Her youngest child was three years old when she conceived again, after she had failed to observe local traditions of celibacy while breastfeeding – her husband had threatened her with a knife and had beaten her when she tried to resist his approaches. Lardi did not want this new infant; she felt that her husband was not a good provider, and complained about him drinking alcohol while she, her co-wives and their children went hungry.
>
> She decided to terminate the pregnancy, and because of the risks involved in illicit terminations for mid-trimester abortions, she was persuaded to go to a hospital. However, the midwife in charge of the maternity ward was related to Lardi's husband, and when she learnt that Lardi was scheduled for an abortion, she abused her for bringing disgrace to the family and refused to admit her without her husband's permission. Lardi returned home, and was assaulted by her husband when she told him of her intention.
>
> The foetus was unharmed and Lardi gave birth to a baby girl – an appropriate punishment, Lardi felt, for her husband's insistence that she carry to term. She subsequently secured a loan from the local women's health co-operative, and had a tubal ligation on a day when her husband's relative was not on duty. She did not inform him about the ligation.
>
> (adapted from Allotey 1996)

The social importance of childbearing is keenly felt by unmarried women, too, despite social controls intended to protect their virginity until marriage. Defo (1997, drawing on Wolf 1973), for example, notes that in the Cameroon, children born out of wedlock are 'joyfully adopted' by the parents of the child's mother. Where unmarried motherhood is not accepted, early pregnancy may be used as a strategy to ensure choice of husband. Premarital pregnancy also provides proof of reproductive ability and so assures a potential partner or his parents of the value of the union. Other emotional and economic reasons also influence very young women's desires to have children (in Haiti, see Maynard-Tucker 1996; among young Nicaraguan women, Berglund *et al.* 1997; for industrialised societies, see Jones *et al.* 1985). Where sexual purity is strongly valued, where it is not possible to adopt the new infant into the mother's family, or where marriage is out of the question, women may proceed with the pregnancy without social support or adequate health care (Bledsloe 1990, Carter 1995) or may choose to avoid stigma and discrimination by seeking an abortion.

In societies where very young women marry or commence sexual relations, childbearing places considerable demands on their health. These pressures are compounded by the physiological stresses of maturation and, in poor communities particularly, by poor nutritional status, anaemia and other health problems as a result of endemic diseases (particularly helminthic infections and malaria), and other infections (Brabin and Brabin 1992). In parts of Sahelian Africa especially, young women's reproductive health is also compromised by genital mutilation (female circumcision) and by reproductive and urinary tract infections. These practises affect the viability of pregnancy and the success of delivery, cause cumulative and chronic health problems (Lightfoot-Klein 1989, Heise 1993, Hicks 1993), and compromise women's sexual health and capacity to enjoy sex.

Jennaway (1996) describes the contradictions for young Balinese women who must balance protecting their reputations with demonstrating their fertility in order to enhance their chances of marrying. Recent research conducted in the Solomon Islands (Burslem *et al.* 1997), too, highlights the value of reproduction as reflected by the family's willingness to care for unmarried pregnant daughters, despite their apparent denial of prenuptial sexual activity and emphasis on the importance of virginity. As suggested, despite the fact that prenuptial sexual relations are often commonplace, public ideologies conventionally hold that unmarried women are not sexually active and consequently young women tend to be poorly educated or misinformed about sexual matters. Berglund *et al.* (1997:9), for example, relate young women's low use of contraception in Nicaragua to misinformation regarding the negative health effects of artificial family planning methods and to religious values which describe conception as 'God's will'. Such ideologies also influence official policies regarding women's access to information, counselling and technologies for fertility control and sexual health. In the Solomon Islands, for instance, government and non-government services are offered only to married women on the supposition of the chastity of unmarried women. Lack of education provided within the community and in schools means that young women have a poor understanding of reproduction, contraception and prevention of infection (Burslem *et al.* 1997).

Reproductive tract infections

Women generally lack information regarding the 'normal' body, and how to manage and treat health problems. Hence women's forbearance of prolapsis, lesions, miscarriages and continence problems. Women's reluctance to report odorous vaginal discharges or other menstrual and sexual problems, however, is due not to their lack of awareness of such problems. Maxine Whittaker (personal communication), in her research on reproductive tract infections, has identified some 60 common terms which Vietnamese women use to describe vaginal discharges, based on differences in colour, consistency, odour and other characteristics such as itchiness. Elsewhere, women may use a generic term for signs of reproductive tract infections (*keputihan* – whitish substance – in Java, Hull *et al.* 1996), and may regard these symptoms as normal or dismiss them as the result of 'stress', 'too many thoughts' (*banyak pikiran*), or 'fatigue' (Hull *et al.* 1996: 233).

The World Health Organization has estimated that there are over 250 million new cases of selected RTIs (reproductive tract infections) and STDs (sexually transmitted diseases) per annum which present a serious problem in terms of increased risk of HIV infection, ectopic pregnancy, miscarriage, neonatal health problems, infertility and cervical cancer (Germain *et al.* 1992). Young women are particularly vulnerable (Lindegren *et al.* 1994, Brabin 1996). The STDs include gonorrhoea, chlamydia, trichomoniasis, syphilis, chancroid, genital herpes, human papilloma virus, and HIV; other RTIs include non-sexually transmitted infections such as candidiasis or bacterial vaginosis, which are sometimes due simply to poor living conditions or other environmental factors, but also to unsafe conditions of childbirth, abortion, IUD insertion, vaginal examination or related procedures, or the insertion of foreign objects or substances to self-treat discharges and irritation (Wasserheit 1990).

RTIs have been consistently neglected by health systems and policy makers (Wasserheit 1990, Aitken and Reichenbach 1994); hence their label as a 'silent tragedy' (Fathalla 1994). Their incidence is higher among women than men, as for physiological reasons they are more likely to acquire such infections, and to suffer more serious health and social consequences. However, while RTIs in men are readily diagnosed and treated, they may be asymptomatic and difficult to identify in women. Women also tend to under-report symptomatic RTIs and are discouraged from seeking advice from health centres, because they regard the symptoms either as 'normal' or 'unimportant'. Or, they avoid reporting such symptoms because of the fear of being stigmatised as 'dirty' or promiscuous (Dixon-Mueller and Wasserheit 1991, Germain *et al.* 1992, Whittaker and Larson 1996). In their study located in Java, Hull *et al.* (1996: 233) note that the women in their study did not perceive long-term or even short-term serious effects of RTI symptoms, were embarrassed by such symptoms, and preferred to self-treat rather than present to a clinic. Self-treatment is common for urinary and sexual health problems. Women routinely self-manage urinary incontinence, for example, constructing the problem as an expected consequence of ageing (Mitteness and Barker 1995) or as the 'normal' outcome of childbearing, menopause and female anatomy (Peake *et al.* in press). Health service providers may share these beliefs and often have limited skills to identify RTIs, lacking knowledge of protocol for case management, counselling and partner notification.

The issues that emerge with respect to RTIs and STDs highlight the ways in which gender, class and ethnicity affect the quality of care within clinical settings. Andrea Whittaker's work (1996a) in northern Thailand among Isan (ethnic Lao) women, for instance, documents their discomfort with the lack of privacy they are given in clinics, their embarrassment at having vaginal examinations, and the dismissive attitudes to which they are often subjected because of their ethnicity and rural background. According to some doctors, village women get reproductive and sexual health problems 'because they are not clean and their hygiene is poor' (Whittaker 1996b: 221). These attitudes are replicated in official government documents (Whittaker 1996b). The alignment of reproductive and sexual health with dirtiness and promiscuity in the minds of women clients and health providers is a major factor discouraging presentation. Similarly, Pike's work in Nepal among young *badi* women illustrates the importance of caste in pre-determining occupation (as sex workers) and in enforcing women's subject status, with obvious relevance to the sexual health risks they face, the quality of health care they receive, and their access to sexual health information (Pike 1997).

Discriminatory attitudes that use ethnicity, language, education and dress as signs of moral worth rather than economic status are not unique to this region. They are commonplace in Australia, and confirm for indigenous women their distrust of white health workers which derives from a long history of institutionalised racism and wilful neglect. This distrust is compounded by the inability of most services to accommodate cultural understandings of 'women's business', which have important implications for the diagnosis and treatment of RTIs, cancer screening, and reproductive health care. Recent research conducted among indigenous women on cervical cancer screening and treatment, for example, highlighted the importance of having examinations conducted by female rather than male doctors, the need for providing clear information about frequency of tests, and the necessity of acknowledging women's fear and embarrassment of the Pap (cervical) smear procedure. As one woman explained:

> I feel very reluctant to go for a Pap smear because of the simple fact that I really don't like going to a clinic. I guess it's the cold and impersonal attitude of the staff that are there. Sometimes you just get the feeling or the sense that this is like an assembly line process . . . It's like that thing where, well it's an invasive thing, an invasive procedure. . . . There is also that other thing too, about the cultural gap, the lack of cultural sensitivity, depending on who the person was, who the technician was who was carrying out the procedure, in some respects it's like being raped . . . I hate even going up for the prenatal check.
>
> (Manderson *et al.* unpublished field notes)

This woman drew attention to the aloofness and distance in clinics, her sense that she was 'a piece of meat' for whom health providers had neither empathy nor respect. Other women also drew attention to the fact that Pap smears and other clinical encounters recalled for women their experiences of rape, sexual abuse and incest.

Age and marital status are barriers to the effectiveness of young health workers to discuss with older, married women sexual, genital, urinary tract and related health problems. Janice Reid (1984), for example, describes the embarrassment of offering and receiving family planning advice between young, unmarried, urban health workers and older, married rural women in Papua New Guinea. Chirawatkul (1993) similarly describes her own reluctance, as a researcher, to talk with older fellow Thai women about menopause and sexuality, in contrast with their earthy humour and openness about sexual health and activity in the focus groups which she organised.

Social issues may also affect the provision of men's sexual health services, but there are fewer data on men's reproductive health. General perceptions are that men are more likely to present for treatment, but cultural perceptions of successful treatment, on the one hand, and kinship relations on the other, may influence this outcome. For example, A<u>n</u>angu Pitjantjatjara in central Australia will avoid treatment for STDs from kin or other familiar community members because of specific kinship relationships and of the fear of gossip regarding the suspicion of sexual and social breaches in behaviour. This makes it difficult to provide sufficient local health workers to meet the needs of all community members in small and isolated communities (Mulvey and Manderson 1995, Willis 1997).

Prevention of HIV

The epidemics of HIV/AIDS have forced academics and health providers to consider how gender and sexuality influence risks of infection. They are also important factors to take into account in developing sustainable behavioural interventions to reduce transmission: sexual health is directly linked to when, with whom and under what circumstances women have sex. Research in industrialised settings and elsewhere indicates that women will not insist on safe sex where other personal reasons militate against this (Catania *et al.* 1992) or where there is inequality between partners (Worth 1989), with gender differences affecting sexual negotiation in various cultural settings and contexts (Johnson *et al.* 1992, Schoepf 1992, Zimet *et al.* 1992, Burslem *et al.* 1997). Attention to gender in this research has highlighted the social contexts and meanings which women invest in sexual relationships, the distinctions that are made where sex is used as a survival or economic strategy within and outside of affective relationships (e.g. de Zalduondo 1991, Muecke 1992), and the use of sex as a means of violence rather than pleasure (Cattell 1996: 99–100, Fischbach and Donnelly 1996).

Women's subordination to men economically and sexually, and the importance of having a partner, mean that women are often willing to accept their partners' statements of faithfulness at face value, and this expectation of trust contributes to women's reluctance to request and insist upon safer sex practices. While women may understand the risk of infection from unsafe sex, economic responsibilities to others in the family also may force them to make different decisions. For example, women who fear desertion or women who are paid for sex are in a weak position to bargain with men to get them to practice safe sex. In addition, women and men both frequently claim to be able to choose 'clean' and 'healthy' sexual partners, and believe that infection can be visually detected on the external body. Choice of partners and preparedness (or not) to use condoms by men as well as women is influenced by beliefs that STDs are visible or can be determined by personal familiarity with prospective partners or by their appearance (being clean and well-dressed) (e.g. Brough 1996, Burslem *et al.* 1997, Willis 1997). Socioeconomic changes, particularly urbanisation, have widened the choice of sexual partners, increased the possibility of paid sex, and multiplied the possibilities for infection by sexually transmissible diseases, highlighting the risks discussed above that are associated with denial of sexual activity among young people.

The sexualisation of non-sexual infections

The association of sexually transmissible infections with dirt and immorality operates as a disincentive for diagnosis and treatment, and for people to seek information and advice on sexual health. Poor quality care in clinical settings reinforces women's reluctance to present with gynaecological and sexual health problems, to seek contraceptive advice, and to participate in screening for such afflictions as cervical and breast cancer (Mensch 1993). Patton (1992:107) characterises the way women are reduced to their metonymic status – as vagina and uterus – and this is particularly apt in capturing the public health and medical focus on women in terms of their reproductive capacity and perceived roles as vectors of disease. Even non-sexual health problems may be sexualised, with important implications for both infected individuals and disease transmission.

The sexual division of labour affects exposure and incidence of infection, accidents and non-communicable diseases (Brabin and Brabin 1992). Our knowledge of the distribution of infection and disease by sex and the physical manifestations of infection is poor because such knowledge tends to derive from clinical cases; this leads to considerable under-reporting where women do not present for diagnosis or treatment. Yet the sequelae of untreated infection – particularly for diseases which cause gross disfigurement, such as lymphatic filariasis, leprosy, onchocerciasis (river blindness) and leishmaniasis, compromise women's physical health and have a high social cost, threatening their life options and personal security through rejection, isolation and/or divorce.

The prevalence of lymphatic filariasis has commonly been assumed to be greater in men than women, with only men affected by such genital complications as hydrocele (enlarged scrotum). However, Brabin and Brabin (1992) suggest that these discrepancies can be explained by the differences in the frequency with which men and women report to clinics for treatment, and argue that, in fact, labial and breast involvement appears to be relatively common. Women are able to hide some signs of infection with their clothes until untreated infection resulting in elephantiasis places serious constraints on their physical mobility and social interactions. Women are reluctant to seek advice about swollen labia or enlarged breasts because they feel uncomfortable discussing their bodies and being examined, especially when the clinic attendants are male.

Medical knowledge of and public health interventions for onchocerciasis have been limited similarly because of the lack of attention to the clinical effects of infection on women. Until recently, control programs emphasised ocular damage leading to river blindness among older people, but overlooked earlier stages of infection. Research in Nigeria (Amazigo and Obikeze 1991, Amazigo 1994) provides graphic examples of onchodermatitis among very young women, who suffer terrible pruritis but do all they can to disguise unsightly lesions from their mothers, potential mothers-in-law and husbands in order to avoid spinsterhood or divorce. Once they are married, however, acute itching, swollen labia, and lesions on the vulva make intercourse painful and sometimes impossible, and many women are finally cast out of their marital homes because of infection.

Feldmeier and colleagues (Feldmeier *et al.* 1993, Feldmeier and Krantz 1993, Anyangwe *et al.* 1994) have argued that the prevalence of genital schistosomiasis, too, has been underestimated due to the low number of case reports and to women's preference for disguising symptoms because they fear that they have a sexually transmitted infection, or that others (health workers and/or kin) will believe them to be so infected. In addition, strong taboos inhibit women's familiarity with their own genitals and bodily processes, impeding their ability to recognise signs of infection. Not all women are able to identify the source of blood (particularly where some genital mutilation has occurred), and not all women notice minor bleeding (urinating and bathing under clothes prevents the most obvious means). Many women suffer various reproductive health problems and urinary tract infections which makes it unlikely that they can distinguish between parasitic infections such as schistosomiasis and other causes of bleeding or pain.

These infections occur primarily in poor populations where young women have little independence from older kinswomen and men; in patriarchal households, mothers-in-law hold both the power and the economic resources to allow younger women to go to a clinic for examination and treatment. Women are especially reluctant to report urinary or genital problems to mothers-in-law or husbands because of fear of accusations of faithlessness. Even illnesses unrelated to the reproductive tract can be sexualised. Difficulties associated with multidrug therapy (MDT) for leprosy in India provide an example of this phenomenon:

> The small dapsone tablets had to be taken in a rather rigid regimen that was difficult for women to follow: before going to bed every night and preferably not on an empty stomach. Health workers suggested to women that they use a calendar to mark each time they consumed a tablet, and a 'blister pack' was introduced to simplify adherence to the prescribed regimen. However, some women became suspicious that they were being given contraceptive pills, particularly as one popular brand of contraceptive pills, MALA-D, advertised on television, looked exactly like dapsone. The health workers frequently had to face mothers-in-law who forbade their daughters-in-law to take MDT for this reason, saying that the distribution of the pills was 'a plot' by family planning advocates.
>
> (adapted from Vlassoff *et al.* forthcoming)

In addition, infectious diseases tend to be most prevalent among the very poor, and among those who are isolated from medical and transport services. Gender also plays a role in affecting access to and use of services. Women may be unable to diagnose or seek advice for infections, tend to receive less information about services than men, and may lack the financial resources to cover transport, service and treatment costs. Domestic responsibilities including care of children restrict women's mobility and the time available to seek health care (Leslie 1992).

Gender relations also shape women's interactions with the health system and in countries and communities where strict gender divisions operate women are often reluctant to present to doctors, especially for sexual health problems. In clinical settings as elsewhere, women's bodies are sexualised, and ideas about their sexuality influence the diagnosis and treatment of problems. In general, primary health care workers are more often men than women, creating problems for women who are uncomfortable about being examined by a strange male, not only for sexual and reproductive health problems but for any ailment. In some societies, there are strong proscriptions against a woman being alone with a man other than her husband, making it almost impossible for her to see a male health worker because of assumptions that the consultation will not be chaste. Her clinic attendance may also be construed as a sexual invitation by the health worker.

Disease prevention is also affected by issues of gender and the control of women's sexuality. For example, various interventions are aimed at increasing the identification of infected people to encourage treatment – active case detection for malaria involving house calls is one example. Where the person making house calls is male, especially where women are secluded, illness may not be reported to prevent interaction between the sexes.

Conclusions

Patton (1992), writing on gender and HIV infections, draws attention to the cultural emphasis on childbearing and its role in devaluing women's other contributions to economic, political and social life. This emphasis on childbearing, she argues, has had a negative effect on women's self-esteem and their willingness to care for their own health, placing them at high risk for STDs and HIV infection, for unwanted pregnancy and untreated disease. In addition, as noted above, women are often embarrassed or reluctant to seek clinical advice, especially for sexual or reproductive health problems, and particularly if they are expected to provide a personal history or be examined physically. Cultural understandings of sexuality and sexual experience affect the prevention, diagnosis and treatment of sexually transmitted diseases including HIV/AIDS, but, as noted, affect women's health far more broadly.

There are potential difficulties related to the reporting, diagnosis and treatment of sexual health problems for men as well as for women. Generally, however, the greater laxity towards men's bodies, lesser emphasis on modesty, and greater tolerance of their sexual behaviour means that men are more likely to recognise and seek attention for signs of genital abnormality and infection. In addition, the potential to identify and treat infections in men is enhanced by their better education and their access to health services. The social control of women's sexuality, and the consequent constraints on their mobility and autonomy, inhibit treatment for sexual, reproductive and non-sexual infectious disease and for other diseases as well, such that all aspects of women's health and well-being are severely compromised.

References

Aitken, I. and Reichenbach, L. (1994). Reproductive and sexual health services: expanding access and enhancing quality. In *Population Policies Reconsidered: Health, Empowerment, and Rights*, ed. G. Sen, A. Germain and L. Chen, pp. 177–92. Boston: Harvard School of Public Health.

Allotey, P. (1997). Researcher or friend? *International Health and Infectious Diseases Study Group Newsletter*, **2**, 1.

Amazigo, U. (1994). Gender and tropical diseases in Nigeria: a neglected dimension. In *Gender, Health, and Sustainable Development*, ed. P. Wijeyaratne, L. J. Arsenault, J. H. Roberts and J. Kitts, pp. 85–99. Proceedings of a workshop held in Nairobi, Kenya, 5–8 October 1993. Ottawa: IDRC.

Amazigo, U. O. and Obikeze, D. S. (1991). *Socio-Cultural Factors Associated with Prevalence and Intensity of Onchocerciasis and Onchodermatitis among Adolescent Girls in Rural Nigeria*. SER/TDR Project Report. Geneva: WHO.

Anyangwe, S., Njikam, O., Kouemeni, L., Awa, P. and Wansi, E. (1994). Gender issues in the control and prevention of malaria and urinary schistosomiasis in endemic foci in Cameroon. In *Gender, Health, and Sustainable Development*, ed. P. Wijeyaratne, L. J. Arsenault, J. H. Roberts and J. Kitts, pp. 77–84. Proceedings of a workshop held in Nairobi, Kenya, 5–8 October 1993. Ottawa: IDRC.

Berglund, S., Liljestrand, J., Marin, F. de M., Salgado, N. and Zelaya, E. (1997). The background of adolescent pregnancies in Nicaragua: a qualitative approach. *Social Science and Medicine,* **44**, 1–12.

Bledsloe, C. (1990). School girls and the marriage process for Mende girls in Sierre Leone. In *Births and Power: Social Change and the Politics of Reproduction,* ed. W. Penn Handwerker. Boulder: Westview Press.

Bledsloe, C. (1995). Marginal members, children of previous unions in Mende households in Sierra Leone. In *Situating Fertility: Anthropology and Demographic Inquiry*, ed. S. Greenhalgh, pp. 130–53. Cambridge: Cambridge University Press.

Bottero, A. (1991). Consumption by semen loss in India and elsewhere. *Culture, Medicine and Psychiatry,* **15**, 303–20.

Brabin, L. (1996). Providing accessible health care for adolescents with sexually transmitted disease. *Acta Tropica,* **62**, 209–16.

Brabin, L. and Brabin, B. J. (1992). Parasitic infections in women and their consequences, *Advances in Parasitology,* **31**, 1–81.

Brough, M. (1996). 'At risk' youth: AIDS, masculinity and the politics of prevention. PhD dissertation, Department of Anthropology and Sociology, University of Queensland.

Burslem, F., Laohapensang, O., Sauvarin, J. and Young, M. (1997). Naked wire and naked truths: Reproductive health risks faced by teenage girls in Honiara, Solomon Islands. PhD dissertation, Master of Tropical Health. Brisbane: Australian Centre for International and Tropical Health and Nutrition.

Carter, A. T. (1995). Agency and fertility: For an ethnography of practice. In *Situating Fertility: Anthropological and Demographic Inquiry*, ed. S. Greenhalgh, pp. 55–85. Cambridge: Cambridge University Press.

Catania, J. A., Coates, T. J., Stall, R. *et al.* (1992). Prevalence of AIDS related risk factors and condom use in the United States. *Science,* **258**, 1101–6.

Cattell, M. G. (1996). Gender, aging, and health: a comparative approach. In *Gender and Health: An International Perspective*, ed. C. Sargent and C. Brettell, pp. 97–122. Upper Saddle River, NJ: Prentice-Hall.

Chirawatkul, S. (1993). Sud Lyad, Sud Luuk: The Social Construction of Menopause in Northeast Thailand. PhD dissertation, Tropical Health Program, University of Queensland.

Defo, B. K. (1997). Effects of socioeconomic disadvantage and women's status on women's health in Cameroon. *Social Science and Medicine,* **44**, 1023–42.

de Zalduondo, B. O. (1991). Prostitution viewed cross-culturally: toward recontextualizing sex work in AIDS intervention research. *Journal of Sex Research,* **28**, 223–48.

Dixon-Mueller, R. and Wasserheit, J. (eds.) (1991). *The Culture of Silence: Reproductive Tract Infection Among Women in the Third World*. New York: International Women's Health Coalition.

Edwards, J. W. (1985). Indigenous koro, a genital retraction syndrome of Southeast Asia: a critical review. In *The Culture-Bound Syndromes: Folk Illnesses of Psychiatric and Anthropological Interest*, ed. R. C. Simons and C. C. Hughes, pp. 169–91. Boston: D. Reidel Publishing Co.

Fathalla, M. (1994). Fertility control technology: a woman-centred approach to research. In *Population Policies Reconsidered: Health, Empowerment, and Rights*, ed. G. Sen, A. Germain and L. Chen, pp. 223–34. Boston: Harvard School of Public Health.

Fischbach, R. L. and Donnelly, E. (1996). Domestic violence against women: a contemporary issue in international health. In *Society, Health, and Disease: Transcultural Perspectives*. ed. J. Subedi and E. B. Gallagher, pp. 316–45. Upper Saddle River, NJ: Prentice Hall.

Germain, A., Holmes, K. K., Piot, P. and Wasserheit, J. (1992). *Reproductive Tract Infections*. New York: Plenum.

Greenhalgh, S. (ed.) (1995). *Situating Fertility: Anthropological and Demographic Inquiry*. Cambridge: Cambridge University Press.

Handwerker, W. P. (ed.) (1990). *Births and Power: Social Change and the Politics of Reproduction*. Boulder: Westview Press.

Heise, L. (1993). Violence against women: the missing agenda. In *The Health of Women: A Global Perspective*, ed. M. Koblinsky, J. Timyan and J. Gay, pp. 171–85. Boulder: Westview Press.

Herdt, G. (ed.) (1982). *Rituals of Manhood: Male Initiation in Papua New Guinea*. Berkeley: University of California Press.

Hicks, E. K. (1993). *Infibulation: Female Mutilation in Islamic Northeastern Africa*. New Brunswick, NJ: Transactions.

Huang, Y. and Manderson, L. (1992). Schistosomiasis and the social patterning of infection. *Acta Tropica*, **51**, 175–94.

Hull, V. J., Widyantoro, N. and Fetters, T. (1996). 'No problem': reproductive tract infections in Indonesia. In *Maternity and Reproductive Health in Asian Societies*, ed. P. L. Rice and L. Manderson, pp. 227–46. Chur, Switzerland: Harwood Academic Press.

Jennaway, M. (1996). Of blood and foetuses: female fertility and women's reproductive health in a North Balinese village. In *Maternity and Reproductive Health in Asian Societies*, ed. P. L. Rice and L. Manderson, pp. 37–60. Chur, Switzerland: Harwood Academic Press.

Johnson, E. H., Gant, L. M., Hinkle, Y. A., Gilbert, D. C., Willis, C. and Hoopwood, T. (1992). Do African-American men and women differ in their knowledge about AIDS, attitude about condoms, and sexual behavior. *Journal of the National Medical Association*, **84**, 49–64.

Jones, E., Forrest, D., Goldman, N., Henshaw, S. K., Rosoff, J. and Wulf, D.

(1985). Teenage pregnancy in developed countries: determinants and policy implications. *Family Planning Perspectives,* **17**, 53–63.
Khanna, S. K. (1997). Traditions and reproductive technology in an urbanizing north Indian village. *Social Science and Medicine,* **44**, 171–80.
Leslie, J. (1992). Women's time and the use of health services. *IDS Bulletin,* **23**, 4–7.
Lightfoot-Klein, H. (1989). *Prisoners of Ritual: An Odyssey into Female Genital Circumcision in Africa.* Binghamton, NY: Harrington Park Press.
Lindegren, M. L., Hanson, C., Miller, K., Byers, R. H. and Onorato, I. (1994). Epidemiology of human immune deficiency virus infection in adolescents, United States. *Pediatric Infectious Diseases Journal,* **13**, 525–35.
Lyttleton, C. (1994). Knowledge and meaning: the AIDS education campaign in rural northeast Thailand. *Social Science and Medicine,* **38**, 135–46.
Maynard-Tucker, G. (1996). Haiti: unions, fertility and the quest for survival. *Social Science and Medicine,* **43**, 1379–87.
Mensch, B. (1993). Quality of care: a neglected dimension. In *The Health of Women: A Global Perspective,* ed. M. Koblinsky, J. Timyan and J. Gay, pp. 235–53. Boulder: Westview Press.
Mitteness, L. S. and Barker, J. C. (1995). Stigmatizing a 'normal' condition: urinary incontinence in late life. *Medical Anthropology Quarterly,* **9**, 188–210.
Muecke, M. A. (1992). Mother sold food, daughter sells her body: the cultural continuity of prostitution. *Social Science and Medicine,* **35**, 891–901.
Mulvey, G. R. and Manderson, L. (1995). Contact tracing and sexually transmitted disease: perspectives of Aboriginal men on the A*n*angu Pitjantjatjara lands. *Australian Journal of Public Health,* **19**, 596–602.
Ojanuga, D. N. and Gilbert, C. (1992). Women's access to health care in developing countries. *Social Science and Medicine,* **35**, 613–17.
Patton, C. (1992). *Last Served? Gendering the HIV Pandemic.* London: Taylor and Francis.
Peake, S., Manderson, L. and Potts, H. (in press). "Part and parcel of being a woman": Constructions of control and female urinary incontinence. *Medical Anthropology Quarterly.* In press.
Pick, W. M. and Obermeyer, C. M. (1996). Urbanisation, household composition and the reproductive health of women in a South African city. *Social Science and Medicine,* **43**, 1431–41.
Puentes-Markides, C. (1992). Women and access to health care. *Social Science and Medicine,* **35,** 619–26.
Ram, K. (1991). *Mukkavar Women: Gender, Hegemony and Capitalist Transformation in a South Indian Fishing Community.* Sydney: Allen and Unwin.
Reid, J. C. (1984). The role of maternal and child health clinics in education and prevention: a case study from Papua New Guinea. *Social Science and Medicine,* **19**, 291–303.

Rice, P. L. and Manderson, L. (eds.) (1996). *Maternity and Reproductive Health in Asian Societies*. Chur, Switzerland: Harwood Academic Press.

Rosaldo, M. and Lamphere, L. (eds.) (1974). *Woman, Culture and Society*. Stanford: Stanford University Press.

Schoepf, B. G. (1992). Women at risk: case studies from Zaire. In *The Time of AIDS: Social Analysis, Theory and Method*, ed. G. Herdt and S. Lindenbaum, pp. 259–86. Newbury Park: Sage.

Sukanya, H. (1988). Prostitution in Thailand. In *Development and Displacement: Women in Southeast Asia*, ed. G. Chandler, N. Sullivan and J. Branson, pp. 115–36. Clayton: Centre for Southeast Asian Studies, Monash University.

Symonds, P. V. (1996). Journey to the land of light: birth among Hmong women. In *Maternity and Reproductive Health in Asian Societies*, ed. P. L. Rice and L. Manderson, pp. 103–24. Chur, Switzerland: Harwood Academic Press.

Vlassoff, C., Khot, S. and Rao, S. (in press). Double jeopardy: Women and leprosy in India. In *Work, Health and Contraception from Women's Perspectives*, ed. M. E. Khan. Baroda: Centre for Operations Research and Training.

Wasserheit, J. (1990). Reproductive tract infections. In *Special Challenges in Third World Women's Health*, ed. The International Women's Health Coalition, pp. 1–15. New York: IWHC.

Whittaker, A. (1996a). Quality of care for women in Northeast Thailand: intersections of class, gender and ethnicity. *Health Care for Women International*, **17**, 435–77.

Whittaker, A. (1996b). White blood and falling wombs: ethnogynaecology in northeast Thailand. In *Maternity and Reproductive Health in Asian Societies*, ed. P. L. Rice and L. Manderson, pp. 207–25. Chur, Switzerland: Harwood Academic Press.

Whittaker, M. and Larson, A. (1996). Reproductive tract infections: the forgotten and neglected component of family planning services. *Venereology*, **9**, 40–7.

Willis, J. (1997). Ritual, romance and risk: Pitjantjatjara masculinity and the prevention of sexually transmissable diseases. Unpublished PhD dissertation, Tropical Health Program, University of Queensland.

Worth, D. (1989). Sexual decision-making and AIDS: why condom promotion among vulnerable women is likely to fail. *Studies in Family Planning*, **20**, 297–307.

Yap, Pow Meng. (1977). The culture-bound reactive syndromes. In *Culture, Disease and Healing*, ed. D. Landy, pp. 340–9. New York: Macmillan.

Zimet, G. D., Bunch, D. L., Anglin, T. M., Lazebnik, R., Williams, P. and Krowchuk, D. P. (1992). Relationship of AIDS related attitudes to sexual behavior changes in adolescents. *Journal of Adolescent Health*, **13**, 493–8.

6

Poverty and the medicalisation of motherhood

SUSAN BRIN HYATT

Medicalisation and modernity

Medicalisation has become one of the primary metaphors through which many contemporary social phenomena are understood and addressed. By 'medicalisation,' I mean the translation of the appearances, behaviours, or 'lifestyles' of individuals occupying particular niches defined as socially problematic into the language of medicine. Through this process, troublesome populations become transformed from living testimonials to the possible deficiencies inherent in the political economy of contemporary life, into 'patients' in need of therapeutic treatment. Both Hopper (1988) and Mathieu (1993), for example, have illustrated how the increase in homelessness in New York City in the 1980s was explained as the result of the deinstitutionalisation of long-term mental patients and of homeless individuals' drug and alcohol addictions, rather than as a reflection of the loss of affordable housing for those now relegated to the margins of the emergent global economy. Each of these authors demonstrates that the re-interpretation of social distress as evidence of individual pathologies, representations which have dominated the press and resultant public policy debates about the increase in homelessness in the United States, have served to mask a much greater problem: that is, the deleterious social consequences of the changing economies of urban spaces.

This logic of linking behaviours seen as socially problematic to putative biological abnormalities has attained a great deal of currency in public discourse. As Urla and Terry (1995:1) note, '[T]he notion that individuals identified as socially deviant are somatically different from 'normal' people is a peculiarly recurring idea that is deeply rooted in Western scientific and popular thought'. While the example above shows how a socially deviant population, the homeless, can be transformed into a *medical* problem, medicalisation as a mode of social analysis has also influenced more broadly the ways in which Westerners have understood the nature of the differences between the sexes. Again, as Urla and Terry (1995:4) have written: '[T]he ideal human body has been cast implicitly in the image of the robust, European, heterosexual gentleman, an ideal configured by its contradistinction to a potpourri of "deviant" types'.

With the male, European, heterosexual, bourgeois body serving as the unmarked norm against which all other populations are measured, then, the female body by definition becomes both socially and biologically anomalous. The programmatic and policy effects of this view of the female have been multiple. Several authors, including Littlewood (1994, Chapter 8), have examined the link between gender inequalities and our understanding of female psychology, showing how women's experiences of stressful social conditions are often misinterpreted as a maladjustment to be regarded as fundamentally psychiatric in nature. In this chapter, I wish to take a slightly different tack and to consider not just the emergence of this view of women as a psychologically problematic population, but to look more specifically at one group whose presumed deviance from a variety of norms was perceived as a danger, not only to themselves but also to the overall stamina of the population as a whole: that is, poor and working-class *mothers*.

Since the eighteenth century, it has been the bodies of women that have been most subject to the clinical gaze of a variety of diagnosticians. Beginning in this period, *mothers* in particular were thrust into a new prominence by a shift from the pre-modern view of the family as a model *for* government to a view of the family as an instrument *of* government (Foucault 1991:100). Now, it was mothers who were held primarily responsible for the successful socialisation of their children; where abnormal behaviour, ill health or infant mortality persisted, the blame for these negative outcomes was increasingly laid at their feet. 'Good' mothering was the panacea for growing social and economic inequality; 'bad' mothering, on the other hand, was believed to foster criminality and other social pathologies.

In earlier periods, differences of position, role, and general well-being among individuals had been explained as manifestations of an inexplicable and inviolable divine will. The contrasting view that emerged through the eighteenth and nineteenth centuries, however, was that society, with all its variation, was neither natural nor inevitable. It was, rather, a domain which could be purposively planned and governed through the judicious application of new technologies such as urban planning, public health, state education, housing design and policing.

It was the development of that complex of technologies designed and deployed toward the end of engineering a 'better' society that readers of Foucault now refer to as 'the social' (Horn 1988, 1994, Rabinow 1989). Mothers became the leading protagonists in this new venture intended to reconfigure society by means of regulation rather than by coercion, and their conduct and compliance was crucial to the overall success of that undertaking. Through such innovations as health visiting and social work, 'women's everyday practices', as anthropologist David Horn (1994:12) puts it, 'were privileged as sites of surveillance and intervention'.

Agendas aimed at social reform became inextricably linked with an emphasis on the importance of creating environments intended to foster 'good' mothering. Campaigns for improving the health and housing of the working classes in Britain, for example, which took on a particular urgency following the devastating losses of World War I, reinforced that connection.[1]

Rather than looking at the differences between men and women's lived experiences of health and illness, then, my intention is to disinter the cultural logic by which women, and poor mothers in particular, became the primary objects of medicalised discourses in the first place. Certainly, the male body has also been a target of medical interventions. I will argue, however, that in the case of mothers, the combination of maternal status and poverty *alone*, without corroborating evidence of any other serious behavioural or psychological dysfunction, has been enough to warrant intensive surveillance and direct intervention. Men, on the other hand, are medicalised not as males *per se,* but, rather, as a consequence of their having been classified as 'deviant' based on their behaviour or lifestyle: as anti-social, homeless, violent, criminally inclined, alcoholic, drug-abusing and the like.

I will begin with the story of one woman's particularly disturbing and gruesome experience of psychosurgery and will historicise this event by looking back at the ways in which a host of measures based on medicine, hygiene and public health have located the causes and cures for social disorder within the bodies and practices of individual mothers themselves, rather than within the political economy of industrial capitalism. In this way, we can see how such taken-for-granted practices as, for example, health visiting and social work, are founded on the assumption that the problems of impoverished mothers, much like those of the homeless, are essentially *medical* problems, to be addressed through the administration of individualised therapeutic regimes, rather than through more broadly-based reforms of the political and economic order.

The story of a 'deviant' woman

Male voice of a psychiatrist: *'You've had this anger your entire life and you tend to have little control over it . . . '*

Another male voice, the film's narrator, explains: *'A psychiatrist's office in Yorkshire . . . '*

A working-class woman speaks: *'I was in the club one night and something started in me and I just picked up a glass and I just stuck it in this woman's face, you know . . . and I didn't realize afterwards that I'd done it, you know . . . '*

Psychiatrist: *'And, you've attacked people and scratched their faces, haven't you?'*

Woman: *'Yes'*

Psychiatrist: *'[You've] rained blows on them'*

Woman: *'Yes'*

Psychiatrist: *'Has your bad temper and your inability to control it, has this got you in trouble with the police or in the courts at all?'*

Woman: *'Yes, I've had five convictions for assault'*

Psychiatrist: *'You have – well, it seems to me that since you've been on tablets and pills for years and you've never really got this major problem in your life under control, it is quite reasonable, VERY reasonable at this stage, to*

consider an operation designed to tone down very considerably indeed your aggressive impulses, to eliminate your bad temper'
Woman: *'Will this help me?'*
Psychiatrist: *'Yes, I think it will'*
Woman: *'It's just a bit frightening, isn't it?'*
Psychiatrist: *'Well, I suppose all operations are a bit frightening – some people are more nervous than others'*

End of scene. In the next scene, Margaret is shown being wheeled into the operating room.

Narrator as a voice-over: *'Margaret is being prepared for an operation that will change her personality. In order to control her temper, a piece of brain, deep inside her head, will be destroyed. The effect will be irreversible and the outcome uncertain. If Margaret is cured, controlling human behaviour by brain surgery will no longer be in the realms of science fiction, but will become day-to-day reality. This is Margaret's case history before, during and after that surgery'.*

(*A Matter of Life: It's a Little Frightening*, Yorkshire Television, 1976).

The woman who was subjected to this operation in 1976 is named Margaret Chapman. She still lives on a council estate [public sector housing] located on the outskirts of the northern city of Bradford. By the time I first met her in 1991, she had become an well-known local activist and was at the forefront of a number of campaigns her community had courageously waged. Since my doctoral fieldwork focused on examining the critical roles undertaken by poor and working-class women in mobilising and sustaining local level grassroots movements in Britain, Margaret's co-operation quickly became integral to my research. The first time I ever spoke with her she recounted to me the startling story of her brain surgery. Astonishingly, this entire event, including the operation itself, was filmed by Yorkshire Television in 1976 and was broadcast as a documentary that same year. Entitled *'A Matter of Life: It's a Little Frightening'*, the film purports to show how doctors believed that a woman diagnosed as 'unusually violent' could be 'cured' through the 'miracle' of brain surgery.

Today, the story of Margaret Chapman's operation might be simply written off as bad medicine; it might be treated merely as the cautionary tale of one wayward doctor's reckless desire to experiment with a vulnerable human subject. Instead of simply dismissing Margaret's experience of psychosurgery as an example of science gone haywire, however, I would like to suggest that we follow the example of Horn (1995), who proposes that we unsettle the boundaries which we believe mark off 'science' from 'pseudo-science' in order to unearth that cultural logic, peculiar to western modernity, which makes both possible.

In fact, attempts to definitively separate 'bad' science from 'good' have proven to be elusive, just as the line separating the normal individual from that of the deviant has also been rendered somewhat indeterminate (Urla and Terry 1995, Horn 1995). In his ingenious re-examination of the work of the late nineteenth century criminologist Cesare Lombroso, for example, Horn (1995) shows that there was an essential and ineluctable connection between those technologies of the social, like social work, which targeted *all* women, and explorations undertaken toward the end of defining the specific nature of one particular type of deviant, in this case, the 'female offender':

> In the end, what emerged from Lombroso's studies was less the (hoped for) transparent pathology of the female offender than the barely legible *potential* dangerousness of the normal woman. As a result, not only were the criminal woman and the prostitute made objects of new practices of surveillance, prevention, and punishment, but the normal woman was placed at the center of a whole range of modern discourses and technologies . . . which ranged from social medicine to social hygiene to social work.
>
> (Horn 1995:109–10)

A brief look at the history of psychiatry reinforces Horn's point that the line separating the 'normal' woman from her deviant sister remains both porous and vague. For psychiatry, which professes to attend to the needs of the abnormal, the troubled, the neurotic and the insane, claims among those patients whom it treats a disproportionately large percentage of women. As Allen writes, 'The female is not the "special" but the *normal* form of the psychiatric patient' (1986:85; emphasis as in the original).

In short, the particular rationality which led to the development of 'the social' in Britain (as well as in other parts of Western Europe, North America and Australia) was, from its earliest inception, profoundly gendered with men constituting the 'norm' thereby relegating all women, especially mothers, to the category of 'deviants', a diverse catch-all which incorporated a wide range of individuals including homosexuals and criminals as well as colonised and non-European peoples.[2] This worldview found perhaps one of its more extreme expressions in the instance of Margaret's surgery. Understanding how this same rationality has operated in somewhat more benign contexts, however, may help us to see the processes through which certain social groups, in this case poor and working-class mothers, have emerged as the objects of scientific attention and have subsequently become the foci for medicalised interventions.

Poor bodies as objects of knowledge

The rise of 'the social' and its attendant mission of reform made the sciences of society, among them anthropology, possible and, indeed, 'necessary', as Horn (1988:395) has written. Collecting data about subaltern colonies of people was essential in order to identify and understand the nature of those populations and their 'habitats' (Horn 1988:396). It was Charles Booth, author of the massive 17 volume *Life and Labour of the People in London*, who, through the medium of the social survey, found a way to translate descriptive material about the working classes into a mode of 'scientific' representation. Contemptuous of the emotive pitch of public discourse on poverty, for which he held popular writings like the Reverend Andrew Mearn's 1883 penny pamphlet, *The Bitter Cry of Outcast London*, responsible (see Wohl 1970, Topalov 1993:398), Booth set out to quantify poverty by replacing lurid stories about the poor with what he considered to be a more reliable sort of information (Topalov 1993:398). As he put it:

> My object has been to attempt to show the numerical relation which poverty, misery and depravity bear to regular earnings and comparative comfort, and to describe the general conditions under which each class lives.
>
> (Booth 1881:277–8, quoted by Topalov 1993:398)

Such information about the poor was not only of scientific interest for knowledge's sake alone: it was also essential in order to put into place appropriate measures for regulating this population. Booth, himself, connected his massive informational surveys and their accompanying poverty maps with concurrent attempts at social reform (Topalov 1993):

> In attempting this work I had one leading idea: that every social problem, as ordinarily put, must be broken up to be solved or even to be adequately stated.
>
> (Booth 1887:375, quoted by Topalov 1993:420)

Similarly, Edwin Chadwick's 1842 *Report on the Sanitary Condition of the Labouring Population of Great Britain* offered a comprehensive survey of the living conditions of the poor, which identified those features that compromised the health not only of the poor but also of the population at large. The transmission of cholera in urban environments, for instance, was of special concern to the middle and upper classes in this period because in contrast to diseases like typhus which remained largely endemic within poor communities, as a water-borne disease, cholera travelled easily outside the confines of the slum (see Tagg 1993:123; Rosen 1976:636–7). As one doctor whose testimony is included in Chadwick's massive report wrote,

> It is not these unfortunate creatures only who choose this centre of disease for their living-place who are affected; but this whole town is more or less deteriorated by its vicinity to this pestilential mass, where the generation of those elements of disease and death is constantly going on.
>
> (Chadwick 1965 [1842]:80)

Statistical surveys of poor neighbourhoods were indispensable for measuring the level of risk they posed to society as a whole and for establishing appropriate 'norms' for hygiene, for personal conduct, and for the correct organisation of living space. If only *everyone* could be made to conform to those norms, the thinking went, such dangers to the population at large could be permanently eliminated (see Rabinow 1989).

The means of ensuring compliance with the guidelines intended to promote healthy environments that developed out of those nineteenth century studies, like those of Chadwick and Booth, included the construction of municipal water and sanitation systems and the passage of housing ordinances determining appropriate population densities. They resulted in schemes for clearing slums, for erecting new housing and for the growth of 'planned' communities. They established a 'scientific' basis for the appointment of agents of enforcement like public health inspectors and social welfare officers. In this way, 'the social' became the project of a vastly expanded public sector, administered by both centralised and local bodies, who employed whole new categories of professionals whose access to knowledge and to state-certified forms of expertise authorised them to govern others, poor and working-class mothers in particular (see Rose 1993:285).

Movements to tackle problems of ill-health among the poor relied on the notion that there was an intimate link between degraded physical environments and the moral and bodily degeneration of those 'unfortunates' who inhabited such surroundings (Ittmann 1995:162). Even while it was deemed essential to educate mothers in their own homes or in clinics and special schools devoted to 'mothercraft',[3] there was also the simultaneous awareness of the possibilities inherent in implementing new technologies intended to engineer a more sanitary and healthful urban *milieu* (Rabinow 1989).

Reform movements thus began to operate on two planes, which sometimes but not always intersected: the urban environment at large became 'a more abstract space – a social–technical environment – upon which specialists would regulate operational transformations' (Rabinow 1989:320) while at the same time, individual mothers also continued to be targets for state interventions. Foucault refers to these two levels of intervention as 'poles'. At one end of this spectrum are those interventions which he refers to as '*regulatory controls: a biopolitics of the population*' (Foucault 1990 [1978]:139). The other pole encompassed a focus on the individual through 'an *anatomo-politics of the human body*' (Foucault 1990 [1978]:139). Science was deployed in the government of each pole towards the end of producing both healthier neighbourhoods and more robust individuals (see Osborne 1996; Rose 1989:128). The flurry of empirical studies of the poor in the nineteenth century, including the extensive poverty maps and surveys of Charles Booth and Edwin Chadwick's 1842 *Sanitary Report*, had distinguished between that environment which was toxic and that which was sound, between contagion-harbouring districts and those judged healthy. At the same time, there was the parallel creation of another new opposition that contrasted 'deviant' individuals with their supposedly 'normal' counterparts.

The depiction of poor mothers as essentially 'deviant subjects', then, who were desperately in need of the normalisation promised by the regimes of the social, was imperative for the successful outcome of all these reforms. After all, the mother was seen not primarily as an individual in her own right but was, rather, as someone who stood in a metonymic relationship to the entire population. She was, at once, the nexus at which her family was connected with 'the race', the conduit through which her children were to be made into productive and healthy citizens of the state (or not), and the juncture at which the unit of her household was conjoined with the institutions of the larger city that surrounded her. If dark and crowded streets fostered fevers and 'miasmas' and needed to be re-designed in order to thwart outbreaks of pestilence, so, too, did the very practices of poor mothers also need to be re-configured as a tactic in the battle against both social disorder and the physical degeneration of the population.

It was mothers who were held responsible not only for the welfare of their own offspring but also for the perpetuation of the British 'race'. In Britain, as in many other European countries at the turn of the century, efforts directed at reforming motherhood took on a new urgency in light of the belief that there was a precipitous rise in rates of infant mortality, a perceived crisis whose cause was frequently attributed to inadequate maternal practices (Dwork 1980:130, Ball and Swedlund 1996[4]). As one of Bradford's crusading social reformers, E. J. Smith, wrote in 1918:

> The declining birth-rate in England and France during the last 40 years, together with the neglect of child life, is in my judgment

> responsible for the War, for had our numbers been maintained, Germany never would have faced the risks involved.
>
> (Smith 1918:65)

In nineteenth and twentieth century England (as in other countries in Europe), however, mothering was hardly considered an endeavour that came naturally to poor women. Even while ideologies of domesticity had begun to permeate Victorian society, holding up 'home and hearth' as the bourgeois woman's proper and desirable sphere of activity (Hall 1992), the poor mother became the primary target of the regimes of the 'social' precisely because she was *not* regarded as innately inclined to be nurturing or caring. Through her supposed inadequacies, she seemed to imperil not only the well-being of her own family but the survival of an entire population.[5]

Those respectable women and men who were bent on reforming inadequate mothers made it clear that it was these harsh realities concerning the prevalence of maternal indifference which necessitated their interventions into family life. In 1908, the Women's Health Association of Great Britain articulated the following position (quoted in Lewis 1980:94): 'There is, indeed, an appalling ignorance of the most elementary principles of domestic hygiene and economy amongst young women of the middle and lower classes'.

In the working class cities of the north of England, like Bradford, many of whose municipal governments in that period were heavily dominated by the social reformist Liberal Party and which had strong nascent Labour Party participation as well, there was a countervailing view of the causes of infant mortality and chronic weakness within the working classes. In this view, infant morbidity and mortality were linked primarily not to maternal practices but to conditions of poverty. At a National Conference on Infant Mortality held in Liverpool in 1914, for example, Dr Helen Campbell, the Medical Officer of Bradford's rather short-lived Municipal Infant Clinic stated that:

> Observation and experience, however, tend increasingly to convince one that the scope of maternal responsibility is strictly limited in that large section of society which chiefly furnishes the need for an infant-life preservation campaign; that it is limited not by maternal

> indifference nor by the existence of maternal ignorance, which can be overcome, but by social and economic conditions which, in the meantime, cannot be. It would seem therefore that *we*, as a nation, must do more *for the infant*, must assume some of the responsibility in order to enable the mother to bear her share.
>
> (Campbell 1914:2; emphasis as in the original)

The case for the involvement of the state in guaranteeing infant welfare was made on the grounds that Britain should be concerned about the quality of its 'racial stock' and future citizens. One headline in the *Yorskhire Post*, for example, confirmed the notion of 'Babies as Municipal Assets' (10 July, 1913). A doctor visiting Bradford in 1919 to see the city's Maternal and Infant Welfare Schemes warned ominously that,

> While we have won the war, we are now losing the race and therefore the peace. We have won at an enormous racial cost. In Bradford you are not even maintaining your numbers. The young population who are to be the future of our country are dying out.
>
> (*Yorkshire Post*, 28 September, 1919)

Whether the causes of infant illness and death were attributed to maternal ignorance, to poverty or to the evils of married women and mothers working in factories (see Smith 1915:70–4, Ball and Swedlund 1996: 40–1), however, the solutions were largely the same. As Dr Campbell (1914:6–7) went on to recommend in her speech to the Conference on Infant Mortality, in addition to the regular monitoring of babies' nutritional status and weight at state-run maternal and infant welfare centres,

> A most essential part of the work of the Infant clinic is the 'following up' or systematic home visitation of the infant by the women inspectors. By a system of exchange of cards, it is possible for the clinic to be kept informed of the results of the visits of the inspectors to the homes, and for those visiting to be kept informed of the observations and advice of the clinic.

The end result was the ever vigilant supervision of impoverished mothers' homes and habits. The project of social reform had been undertaken in the first place supposedly to *include* the poor not to exclude them, to enmesh them within that overarching and totalising web of normalisation. And yet, it was the degree to which poor mothers were assiduously policed, through home visiting and social work, public health and urban planning, that reinforced to them their own apparently immutable status as 'deviants'.

The poor mother had become the embodiment of the pathogenic city which surrounded her, a potential incubator of illness and infant death. Just as technology would now allow for the regulation of the ebb and flow of water, sewage and light in order to refashion the city into a hygienic environment, so did the poor mother also need to be re-made in the image of the 'good' (read: bourgeois) mother, whose health-giving capacities and moralising properties were considered essential for the well-being and betterment, not only of her own family but also of society as a whole.

The medicalised mother

It was the inadequate and impoverished mother, whose capacity to compromise the vigour and future labour of the working-classes of Britain rendered her a latent criminal (see Castel 1991).[6] The reinterpretation of her distress as a physical or psychological disorder, amenable to therapeutic treatment, (Kleinman 1995:36) provided a strategy for the containment of the poor while at the same time reducing the risks of widespread social strife as a consequence of concurrent growing economic inequality.

One indicator of the degree to which this medicalising discourse continues to penetrate the lives of poor women can be seen in their present day high rates of consumption of psychotropic drugs – tranquillisers (benzodiazepines), in particular – which are routinely prescribed by their local GPs (see Littlewood 1994, McDonald 1994:20–1). Such pharmaceutical regimes remove the analysis of a woman's mental state from the total context of the material conditions of her everyday life, allowing for the identification of a series of discrete symptoms located within the internal confines of the woman's own body and psyche.

Among the women on the council estates in Bradford, almost every one whom I interviewed acknowledged using prescription sleeping pills on a regular basis; in many cases, this use was coupled with the taking of amphetamines ('uppers') to be able to be alert for children during the day. Even though these drugs seemed to be readily available through prescriptions from local GPs, they were also bought and sold and were part of an indigenous system of exchange and reciprocity. Often while I was visiting at one woman's house, another woman would come by asking to borrow a 'moggie', a nickname for the drug trade-named Mogadon (generic name: nitrazepam) or some other familiarly prescribed tranquilliser: 'Aw, just loan us a moggie, so I can get some sleep tonight, love', the women would implore one another.[7]

That the women used nicknames (like 'moggies') to refer to these tablets made them seem almost like objects of affection; they were frequently shared around, with women recommending to one another preferred prescriptions and dosages. In this manner, they, along with cigarettes and drinks at the pub, were integrated into a local system of sharing that marked neighbours as friends and allies.[8]

Margaret Chapman's medical history prior to the time of her surgery offers an instructive glimpse into the consequences of the over-medication that is typical of many women in her situation. In her published autobiography, *No Option but to Fight*, Margaret describes how she first began using prescription drugs, in this case, amphetamines. Trapped in a violent marriage, saddled with three small children and pregnant with her fourth by the age of 25, she consulted her doctor:

> I told him that I was feeling worn out and that my mind was willing to do things but my body wouldn't shift [move]. He wrote me a prescription and I went round to the chemist [pharmacist] to collect my 'cure'. Until then I had never taken a tablet in my life. These were a bluey purple colour, heart shaped, and on the bottle was written, 'Two to be taken every morning'. I swallowed a couple, not knowing what effect they would have. A couple of hours later they began to have a stimulating effect and I started to feel great, better than I had felt in years. My mind and body began to function again. I felt as though I had returned from the dead. I was alive again. The grass looked greener and the sky so blue, in fact everything looked brighter. This was a wonder drug.

> . . . I had no idea of how those amphetamines were going to lead to a major crisis in my life.
>
> (Chapman 1990:34–5)

Many of the other women whom I interviewed also acknowledged suffering from a variety of recognised psychological illnesses such as agoraphobia (a term whose meaning they correctly understood), stress, nerves, anxiety and depression. They also complained of attendant symptoms of these conditions including insomnia, fatigue, overeating, anorexia, alcoholism, prescription drug dependency and addiction to smoking.[9]

Given the living conditions that most of these women were contending with, such as sub-standard housing, crime, male violence, lack of childcare and physical isolation, I was not surprised to hear this catalogue of ills. Indeed, the only aspect of these discussions that was surprising was the extent to which the women willingly participated in the somatisation of their life circumstances, reducing external pressures and deficiencies to 'stress' or other medical conditions and attributing their hardships to their own individual 'inability to cope'.

It is too simple – though perhaps not entirely inaccurate – to suggest that doctors treat impoverished women on British council estates with psychotropic medications because that is all they have to offer in the face of circumstances commonly referred to as multiple deprivation.[10] Clearly, it may not be within the doctor's power to provide women with a source of food, with new housing, with jobs, or with child care. Yet, this explanation does not entirely satisfy because it leaves relatively untouched those assumptions underlying that rationality which would place social distress within a language and treatment of bodily causes to begin with.

Intrinsic to the project of social reform, however, was precisely that capacity to break down bodies, environments and other 'problems' into their constituent elements, to divorce their parts from the whole in the attempt to 'know' them and then to be able to subsequently find ways to 'treat' or to 'cure' them. Poverty, for example, had to be represented in a neat and orderly fashion so that it could be addressed in a systematic manner. It needed to be colour-coded, quantified, and reduced to two dimensions through the medium of Charles Booth's visual maps, which located it within particular bounded districts; it was able to be sanitised and made sterile by condensing its casualties into tables of statistical calculations and censuses. As Tagg (1988:135) writes in his analysis of the use of photographs as part of a campaign by doctors in favour of slum clearance in turn-of-the-century Leeds,

> It is a remarkable, if successful, strategy which says nothing of political reversals or the intractability of the slum population to 'regular industry', schooling or moral reform, but speaks only of space, light and air, resting its case for clearance on the technical claims of a pseudo-medical discourse itself underpinned by the technicism of photography.

Photographs, along with street plans, grids and maps transformed the hidden crannies of poor communities into clearly delineated geometrical configurations, amenable to manipulation through the *standardised* application of new technologies. Under the omniscient gaze of the surveying expert, unlit alleys and fetid cubby-holes were made visible and were thus rendered *knowable.* Similarly, under the regimes of medicalisation, the woman is reduced from a whole person into a compendium of signs and symptoms for which the appropriate therapies can be prescribed and administered. Both the poor mother and her debilitating environment are thus neutralised as sites of potential danger, their infectious capacities inactivated through the application of scientifically derived interventions. By pathologising the mental state of the poor mother, then, and by acting upon her body to 'normalise' her through the administration of drugs (the anatomo-politics of the human body), the greater goal of managing the population as a whole (biopolitics) is partially realised.

In cases like that of Margaret's, the search for manifestations of deviance did not end with attempts to modify her living spaces and practices: her very corporeal being came under suspicion. New technologies of science and medicine now allowed for the possibility of investigating the actual bodies of deviants in order to discover the biological reasons for their apparently abnormal behaviours.

The doctors who examined Margaret and, by extension, the

makers of the television documentary cast her biography in the form of a medical 'case history'; every episode in her life that is shown, therefore, appears chosen as if to provide the viewer with irrefutable evidence of her inherent pathology and of the dangers she posed to others, rather than, alternatively, serving to illustrate her possibly quite reasonable responses to the pressures of poverty. As framed by that 'case history' context, scenes in her council house with her children, engaged in activities as innocuous as peeling potatoes, suddenly take on an ominous undertone. One watches these scenes anxiously in search of the tell-tale aberrance that will herald what is deemed to be an irrational outbreak of violence. No such episode occurs and, indeed, in interviews with Margaret's neighbours shown on the film, they all agree that her violence was never directed toward her children but that it was, in fact, often undertaken in defence of their well-being.[11]

Throughout her history of treatment for various ailments including addiction to prescription drugs (which was largely an outcome of the medical treatment she had received), alcoholism and depression, Margaret was periodically judged to be an 'unfit mother' and her children were regularly taken into custody by the state. In fact, according to her own autobiography (Chapman 1990:88–92), Margaret, having become overwhelmed by the demands of her now five teenaged children and with no support system made up of other adults (she had been raised in an orphanage and had no mother or female kin living nearby), asked to admitted into hospital for treatment for her depression. Her children were again taken into care. It was at that point, when she was at her most vulnerable and fearful about the future of her relationship with her children, that she was persuaded to consent to the surgery.[12]

For the surgery, Margaret had to be conscious while the doctor probed through a hole drilled in her skull, looking for the part of her 'hypothalamus'[13] which he claimed had to be 'burned out' in order to rid her of her violent behaviour. She is shown dressed in a hospital gown, her personhood diminished as she is made into the passive object of expert medical interventions, the patient. By the end of the film, the narrator sounds obviously shaken as he questions the surgeon:

Narrator: *'If her violence stemmed from her social background, if she was in a different set of circumstances, she might not have needed the operation'*

Psychiatrist: *'Well, that's a possibility, I suppose. It's a possibility'*

Narrator: *'So it's a possibility that the operation wouldn't have been necessary had she lived in different circumstances'*

Psychiatrist: *'That is possible. I'm not saying it's probable, but it's possible'*

Narrator: *'In other words, she wasn't born to be violent'*

Psychiatrist: *'I don't know'*

Narrator: *'And, so it's desirable, is it, from your point of view, to change all of those people?'*

Psychiatrist: *'It would be desirable for any society to convert its aggressive people, providing you don't interfere, for example, with their intellect, into non-aggressive people. Don't you think that would be a good thing?'*

Narrator: *'So you could go into all of the jails and do the operation on all of the violent prisoners there?'*

Psychiatrist: *'Yes, that would be a very good thing'*

Narrator: *'So that there's no violence in the world at all?'*

Psychiatrist: *'No, no, you're generalising too much that there would be no violence in the world. But, I still maintain that if we could convert by a relatively simple operation such as this, if we could convert a violent criminal into somebody who would fit in, and who would be a normal person, that would be a good thing'*

Narrator: *'Would you like to see that happen then?'*

Psychiatrist: *'I'd like to see attempts made in that way, yes'*

(*A Matter of Life: It's a Little Frightening*, Yorkshire Television, 1976).

What the actual effect of the operation was, I was never able to ascertain. Margaret frequently complained that it left her with severe headaches, an inability to concentrate well, and a slackness of her jaw. Her supposed propensity for violence seemed to dissipate many years later, after her children were grown and she had become involved in grassroots campaigns aimed at bettering conditions in her community. Margaret, herself, attributed her relative calm as she grew older to her having learned to channel her frustrations through political activism rather than as any consequence of the surgery. Indeed, the film, itself, ends ambiguously with Margaret reporting to the doctor on screen that even after the operation, she still felt compelled toward violence.

The story of Margaret's brain surgery may well be atypical; nonetheless, it highlights the assumptions which underlie the ways in which reform projects engage in strategies of normalisation through producing evidence of 'treatable' deviance in socially problematic populations. To be sure, men have also been the targets of such efforts but not solely as a consequence of their sex. The characteristics attributed to *all* poor mothers, however, by practitioners involved in the Infant and Maternal Welfare movement, such as medical officers, health visitors and social workers, left such mothers besmirched by a legacy which placed them always just on the brink of falling into the abyss of the abnormal. Without appropriate surveillance and interventions, they could not be relied upon to care for their children properly, endangering not just their own families but the future of an entire race and nation; the consequences might mean that they should forfeit altogether the right to *be* mothers.

Science and medicine offered the new option of psychosurgery as a particularly invasive and dehumanising means for promoting a harmonious society through defusing the potential for violence and social disruption allegedly inherent within the body of the deviant. Like the slum dwelling, the body had also become decontextualised and isolated, making it and its infrastructure, its biology, its hormones, and even its very brain appropriate sites for the application of biomedical interventions.

As current agendas aimed at social improvement have become ever more therapeutic in nature, intended to reconfigure individuals rather than society as a whole, the association between social problems and physical bodies continues to be applied to a range of 'suspect' populations. Indeed, some current scientific agendas now call for new research to investigate a putative connection between male violence and genetics.[14] Instead of remaining an account of a sensational event now safely locked away in the archives of an obsolete past, Margaret's experience of psychosurgery may actually turn out to have been a frightening harbinger of a 'brave new world' of biomedical and genetic engineering yet to come, which continues to submit that the roots of social unrest remain to be discovered, not within the structure of society, itself, but, rather, lurking amidst errant chromosomes and defective nervous systems concealed deep within the organic corporeality of the individual human body.

Endnotes

1 See Allman (1994) on the training of 'good' mothers as an integral part of colonialism's 'civilising mission'.

2 For discussions of the medicalisation of colonised populations, see J. and J. Comaroff (1992) and Quinlan (1996). Kaw (1993) presents interesting documentation of the medicalisation of racial features and consequent plastic surgery undertaken by Asian American women.

3 Alongside the strategy of visiting new mothers in their own homes, there was also the establishment of infant and baby public health clinics and so-called 'schools for mothers', which became even more popular following the devastating loss of human life resulting from World War I (see Dwork 1987, Dyhouse 1978:249–59, Ross 1993:215–19). According to social historian Ellen Ross (1993:215), by 1920, there were 1583 such clinics and centres located in England and Wales.

4 Ball and Swedlund (1996) present an excellent study which compares attitudes toward infant mortality and prevailing beliefs about the adequacy of mothering practices in Britain and in the US.

5 Davin (1978) has presented the key discussion on the link between Britain's imperialist project and the reform of mothers.

6 In the present political climate, poor mothers continue to be particularly criminalised. See Gomez (1997) for one recent treatment of the issue of drug addiction during pregnancy and the policing of pregnant women and new mothers.

7 Mogadon, along with Valium and Librium, were the most frequent benzodiazepines prescribed in Britain. An audit of hypnotic and anxiolytic use conducted by Bradford Health Authority in September 1993, showed that Mogadon, one of the most addictive of the benzodiazepines, accounted for 24 % of the tranquilliser prescriptions written in a sample of 323 patients spread over 5 practices in Bradford District (Bradford Health Authority unpublished report).

8 This sharing of medications was not only true of tranquillisers; on several occasions I also observed women offering one another oestrogen tablets (hormone replacement therapy) for 'moodiness' and menopausal symptoms and nicotine patches for those gamely trying to quit smoking.

9 Smoking is a complicated phenomenon, because, although women were well aware of the health risks associated with this practice, they also valued it for its social dimensions; smoking was a collective activity in which cigarettes were shared around. It also marked off adult space from children's space (see Graham 1987).

10 See Ettore (1992), Helman (1994:211–26) and Cooperstock and Lennard (1986) for discussions of the social meanings of tranquilliser use among women.

11 Margaret, herself, often pointed out that what seemed to constitute abnormally 'violent' behaviour for a woman would have passed virtually unnoticed in a man.

12 It was clearly easier for the authorities to see Margaret's body as 'abnormal' rather than to attempt to address the issues created by her poverty. As Margaret, herself, once told me, 'Single parents have to be deviants in order to survive. If you go out and steal to feed your kids, they punish you. Yet, you're trying to take care of your kids. If you neglect your kids, they come and take them away from you'.

13 Curiously, abnormalities of the hypothalamus remain sources of scientific 'suspicion' and are now being investigated as a cause of homosexuality – see Terry (1995) for a critique of that research.

14 See Wright (1995) and Gladwell (1997) for popular discussions of the new research linking biology and violence.

Acknowledgements

First and foremost, my greatest thanks are to Margaret Chapman, who has supported my research from the beginning and has been more than generous in sharing the details of her remarkable life story with me. I am also grateful to Tessa Pollard, Helen Ball and Sydney White for their thoughtful readings and helpful comments in response to earlier drafts of this chapter, and to Jacqueline Urla and Susan DiGiacomo for their comments on this material as it originally appeared in my doctoral dissertation. Participants in the 1997 Biosocial Society Annual Workshop were a wonderful audience for the initial presentation of this work, and I thank the Biosocial Society for its support.

The fieldwork on which this article is based was conducted with the help of grants from the Social Science Research Council and the National Science Foundation (DBS-9223510) in the USA. A Faculty Research Grant from Temple University during the summer of 1997 allowed me to spend time working in the West Yorkshire Archives and in the local history collection of Bradford City Library, gathering the historical materials on the Maternal and Infant Welfare Movement in Bradford. I am grateful as well to the staffs of those institutions for their help, especially Carol Greenwood.

Despite all of this excellent assistance, I take full responsibility for this work and for any errors of fact or interpretation.

References

Allen, H. (1986). Psychiatry and the construction of the feminine. In *The Power of Psychiatry*, ed. P. Miller and N. Rose, pp. 85–111. Cambridge: Polity Press.

Allman, J. (1994). Making mothers: missionaries, medical officers and women's work in colonial Asante, 1924–1945. *History Workshop Journal*, **38**, 23–47.

Ball, H. L. and Swedlund, A. (1996). Poor women and bad mothers: placing the blame for turn-of-the-century infant mortality. *Northeast Anthropology*, **52**, 31–52.

Campbell, H. (1914). The provision of medical treatment at infant consultations; or the scope of the infant clinic. *National Conference on Infant Mortality, Liverpool, July 2 and 3*. London: John Bale, Sons & Danielsson, Ltd.

Castel, R. (1991). From dangerousness to risk. In *The Foucault Effect: Studies in Governmentality*, ed. G. Burchell, C. Gordon and P. Miller, pp. 281–98. Chicago: University of Chicago Press.

Chadwick, E. (1965 [1842]). *Report on the Sanitary Condition of the Labouring Population of Great Britain*. Edinburgh: Edinburgh University Press.

Chapman, M. (1990). *No Option But to Fight*. Glasshoughton, West Yorkshire: Yorkshire Arts Circus.

Comaroff, J. L. and Comaroff, J. (1992). Medicine, colonialism and the black body. In *Ethnography and the Historical Imagination*, ed. J. L. Comaroff and J. Comaroff, pp. 215–34. Boulder: Westview Press.

Cooperstock, R. and Lennard, H. L. (1986). Some social meanings of tranquilliser use. In *Tranquillisers: Social, Psychological and Clinical Perspectives*, ed. J. Gabe and P. Williams, pp. 227–43. London: Tavistock.

Davin, A. (1978). Imperialism and motherhood. *History Workshop Journal*, **5**, 9–65.

Donzelot, J. (1979). *The Policing of Families*. New York: Pantheon Books.

Dwork, D. (1980). *War is Good for Babies and Other Young Children: A History of the Infant and Child Welfare Movement in England, 1898–1918*. London: Tavistock Publications.

Dyhouse, C. (1978). Working-class mothers and infant mortality in England, 1895–1914. *Journal of Social History*, **12**, 248–67.

Ettorre, E. (1992). *Women and Substance Use*. Basingstoke: Macmillan.

Foucault, M. (1990 [1978]). *The History of Sexuality: Volume I: An Introduction*. New York: Vintage Books.

Gladwell, M. (1997). Damaged. *The New Yorker*, February 24 and March 3, pp. 132–47.

Gomez, L. E. (1997). *Misconceiving Mothers: Legislators, Prosecutors, and the Politics of Prenatal Drug Exposure*. Philadelphia: Temple University Press.

Graham, H. (1987). Women's smoking and family health. *Social Science and Medicine*, **25**, 47–56.

Hall, C. (1992). The early formation of Victorian domestic ideology. In *White, Male and Middle-Class: Explorations in Feminism and History*, ed. C. Hall, pp. 75–93. Cambridge: Polity Press.

Helman, C. G. (1994). *Culture, Health and Illness: An Introduction for Health Professionals*. Oxford: Butterworth/Heinemann.

Hopper, K. (1988). More than passing strange: homelessness and mental illness in New York City. *American Ethnologist*, **15**, 155–67.

Horn, D. (1988). Welfare, the social and the individual in interwar Italy. *Cultural Anthropology*, **13**, 395–407.

Horn, D. (1994). *Social Bodies: Science, Reproduction, and Italian Modernity*. Princeton, NJ: Princeton University Press.

Horn, D. (1995). This norm which is not one: reading the female body in Lambroso's anthropology. In *Deviant Bodies: Critical Perspectives on Difference in Science and Popular Culture*, ed. J. Terry and J. Urla, pp. 109–28. Bloomington: Indiana University Press.

Ittmann, K. (1995). *Work, Gender and Family in Victorian England*. New York: New York University Press.

Kaw, E. (1993). Medicalization of racial features: Asian American women and cosmetic surgery. *Medical Anthropology Quarterly*, **7**, 74–89.

Kleinman, A. (1995). *Writing at the Margin: Discourse between Anthropology and Medicine*. Berkeley: University of California Press.

Lewis, J. (1980). *The Politics of Motherhood: Child and Maternal Welfare in England, 1900–1939*. London: Croom Helm.

Littlewood, R. (1994). Symptoms, struggles and functions: what does the overdose represent? In *Gender, Drink and Drugs*, ed. M. McDonald, pp. 77–98. Oxford: Berg.

Mathieu, A. (1993). The medicalization of homelessness and the theatre of repression. *Medical Anthropology Quarterly*, **7**, 170–84.

McDonald, M. (1994). Introduction: A social-anthropological view of gender, drink and drugs. In *Gender, Drink and Drugs*, ed. M. McDonald, pp. 1–31. Oxford: Berg.

Mearns, A. (1970[1883]). *The Bitter Cry of Outcast London*. New York: Augustus M. Kelley Publishers.

Osborne, T. (1996). Security and vitality: drains, liberalism and power in the nineteenth century. In *Foucault and Political Reason: Liberalism, Neo-Liberalism and Rationalities of Government*, ed. A. Barry, T. Osborne and N. Rose, pp. 99–121. Chicago: University of Chicago Press.

Quinlan, S. (1996). Colonial encounters: colonial bodies, hygiene and abolitionist politics in eighteenth century France. *History Workshop Journal*, **42**, 106–25.

Rabinow, P. (1989). *French Modern: Norms and Forms of the Social Environment*. Cambridge: MIT Press.

Rose, N. (1989). *Governing the Soul: The Shaping of the Private Self.* London: Routledge.

Rose, N. (1993). Government, authority and expertise in advanced liberalism. *Economy and Society*, **22**, 283–99.

Rosen, G. (1976). Disease, debility and death. In *The Victorian City: Images and Realities*, Volume 2, ed. H. J. Dyos and M. Wolff, pp. 625–67. London: Routledge and Kegan Paul.

Ross, E. (1993). *Love and Toil: Motherhood in Outcast London, 1870–1918.* Oxford: Oxford University Press.

Smith, E.J. (1915). *Maternity and Child Welfare: A Plea for the Little Ones.* London: PS King & Son, Ltd.

Smith, E.J. (1918). *Housing: the Present Opportunity.* London: P.S. King & Son, Ltd.

Tagg, J. (1988). *The Burden of Representation: Essays on Photographies and Histories*, pp. 117–53. Minneapolis: University of Minnesota Press.

Terry, J. (1995). Anxious slippages between 'us' and 'them': a brief history of the scientific search for homosexual bodies. In *Deviant Bodies: Critical Perspectives on Difference in Science and Popular Culture*, ed. J. Terry and J. Urla, pp.129–69. Bloomington: Indiana University Press.

Topalov, C. (1993). The city as *terra incognita*: Charles Booth's poverty survey and the people of London, 1886–1891. *Planning Perspectives*, **8**, 395–425.

Urla, J. and Terry, J. (1995). Introduction: mapping embodied deviance. In *Deviant Bodies: Critical Perspectives on Difference in Science and Popular Culture*, ed. J. Terry and J. Urla, pp. 1–18. Bloomington: Indiana University Press.

Wohl, A. (1970). Introduction to *The Bitter Cry of Outcast London* by the Reverend Andrew Mearns [1883]. Leicester: Leicester University Press.

Wright, R. (1995). The biology of violence. *The New Yorker*, March 13, pp. 68–77.

7

The vanishing woman: gender and population health

PATRICIA A. KAUFERT

Franklin (1997) describes anthropologists as fascinated by the 'presence of an absence', particularly when the anthropologist is a woman and the absent figure is female. Traditional ethnographers constructed their complex and elaborate models of political and economic systems, religion, law or kinship based largely on conversations with men. Insofar as they made an appearance in these models, women were usually figures with children, passing silently through public space. Passive actors, they were objects in a script created and recounted by men for the visiting anthropologist. Similarly, this chapter is also concerned with models in which women, while present, are invisible; these models, however, are of more recent origin and are constructed by health economists and health policy analysts rather than by ethnographers.

The advent of feminist scholarship disturbed the comfort of the anthropological world, challenging its theoretical assumptions by showing how differently social institutions – such as kinship – appear when seen by, and through the eyes of, women (Strathern 1995). To see from this different perspective, researchers had to enter the 'invisible' world of women and engage with the reality of their lives. Abu-Lughod (1995:22), for example, moved literally and figuratively into the tents of the women, where she discovered that Bedouin women were:

> Living in a separate community – a community that could also be considered a subsociety: separate from and parallel to the men's, yet crosscut by ties to men.

Once she had crossed into their community and had been accepted by them, the women described for her how kinship and family ties operated in practice rather than in theory. Using these insights, Abu-Lughod came to understand Bedouin society much more fully, which would have been impossible had she talked only with the men, accepting unquestioningly their model of how the world worked. Instead, she was able to make Bedouin women visible in ways they would not have been in traditional ethnography. Rather than remaining passive, silent figures, they become active presences, speaking out and commenting on their own lives as well as on the lives of men.

Many women scholars have, like Abu-Lughod, a sense of moral as well as scientific commitment to providing women with visibility and voice, whether they are Bedouin women in Egypt, or women at an IVF (*in vitro* fertilisation) clinic in England (Franklin 1997) or in a prenatal screening programme in New York (Rapp 1997). Rapp, Franklin and Abu-Lughod, all worked in the same ethnographic tradition as many of the other contributors to this volume, listening closely to women and using their voice to shatter the silence around them. This chapter is rooted in the same sense of commitment, but it is dealing with an issue – the relative invisibility of women within the Population Health model – which is a construct of academics and not a 'reality' in the same sense that an amniocentesis clinic or the IVF process is 'real'. As a model, it cannot be directly experienced by women; yet, insofar as it has become a basis for policy making, its consequences for women are real. It is important, therefore, to understand its construction, to identify its flaws and to examine its implications for women.

My intention in writing this chapter is to look at this model from the perspective of the anthropologist, using the insights that come from this discipline to question the way in which epidemiologists think, select and put together the data to construct their version of how the world works. My approach is borrowed from those anthropologists working in social studies in science, who have adopted the laboratory with its scientists as their fieldwork site (Oudshoorn 1997:42). Their interest lies in explaining how a scientific product came into existence; their questions include: 'Who developed this product and why?'; 'What materials were used?'; 'How did material and technique influence design?'; 'How did design determine use?'; 'What is in "the black box"?' Fujimara's brilliant study of proto-oncogene cancer research, for example, opens with a detailed analysis of the development of molecular genetic technologies in the early 1980s. Defining what is meant by 'black-boxing' she writes:

> The term 'black box' refers to a tool that is no longer questioned, examined or viewed as problematic, but is taken for granted.
>
> (Fujimara 1996:213)

Fujimara herself uses the term when describing the 'black-boxing' of a technique for manipulating DNA using restriction enzymes, but a black box may be a set of formulae, a text, a construct or model (Latour 1987, Latour and Woolgar 1986).

Transferring the methodology from the laboratory studied by Fujimara to the very different setting of health policy analysis, I have replaced the genetic scientists with health economists and health policy analysts and have defined their scientific product as a series of texts (books, papers, conferences, special issues of journals) in which they debate and elaborate on their model of population health. The black box is the model itself. My role as an anthropologist is to open up the black box, poke around in its interior, look for the women, and ask 'who hid them, how and why?'

Who are the scientists?

The European and North American debate over the role of the State in relation to the health of the citizenry created a demand for experts and analysts. Starting in the late 1980s, academics and health researchers were invited into the public arena in unusual numbers, presented with the task of finding a solution to escalating health costs. In Canada, the language and the theoretical constructs used in framing the discussion on the future of health care were strongly influenced by the members of the Program on Population Health, a high-powered think tank for health policy makers and academics, set up by the Canadian Institute for Advanced Research (CIAR). A series of papers written by members in this group resulted in a book, *Why are Some People Healthy and Others Not? The Determinants of Health of Populations* (Evans *et al.* 1994), in which the main thesis is that medical care makes only a minor contribution to population health and that the major determinants of health are social and economic. In this model, the well being of the population depends not on medical care, but on a relatively equitable distribution of income, on a social environment which provides people with a sense of security and control, on stable and satisfying employment, and on the availability of social support.

Locating health in the social conditions of people's lives is an idea which can be dated back to the origins of the public health movement and the work of Villerme in France (Krieger 1992) and Virchow in Germany (Rosenberg 1995). Within the past 25 years, the Lalonde Report (1974) in Canada and the Black Report (Black *et al.* 1988) in the UK put forward many of the same arguments. Variations on these themes have been proposed by McKeowan (1976), Navarro (1986) and McKinlay (1993). Yet, despite being neither particularly new nor original, the book made a very rapid transition from being simply another academic publication to becoming the primary text of reference for government and its officials, widely read and debated in the circles of health policy makers and analysts.

Shortly after *Why are Some People Healthy and Others Not?* was published, I was asked to write a paper using the model proposed by Robert Evans and his colleagues, but exploring how the determinants of health were expressed in women's lives (Kaufert 1996). It was an interesting exercise, but its most intriguing aspect was the marked absence from the book itself of any discussion of gender issues. Women seemed to be as invisible in this text as they were in traditional ethnographies, suggesting that this might be yet another case of 'the discursive exclusion of women from cultural representations and social practices' (Ginsburg and Rapp 1995:3). But how had women – a substantial presence in any population – been made to vanish so easily into the black box, as if by some vaudeville magician? I subsequently discovered that rather than using the mirrors and trapdoors of the illusionist, the Population Health group had depended on a form of statistical sleight of hand.

I started to monitor other texts on health policy as they were published over the next few years, looking always for the women. I have restricted the analyses for this chapter to key texts, including, 1. *Why are Some People Healthy and Others Not?* (Evans 1994), 2. a special issue in 1994 of *Daedalus* (the journal of the American Academy of Arts and Sciences) on the topic of 'Health and Wealth', 3. Richard Wilkinson's (1996) *Unhealthy Societies: The Afflictions of Inequality*, and 4. a collection of papers published in the *American Journal of Public Health* in 1997.

The contributors to these texts are members of an international network of scientists. The papers in *Daedalus* were the product of a conference funded by Honda, a Japanese Foundation but with Canadian, Japanese and American contributors. Wilkinson is British Canadian, and both British and American researchers contributed to the *American Journal of Public Health.* Although drawn from different countries and disciplines, the members in this network attend the same conferences, write in the same journals and draw on the same literature.

This is not just a network of scientists but a network of scientists who are all relatively eminent in their fields. Richard Wilkinson is a Senior Research Fellow at the Trafford Centre for Medical Research at the University of Sussex. Among the American members, Theodore Marmot is a Professor of Public Policy at Yale and Leonard Syme is a Professor of Epidemiology at the University of California at Berkeley. The original members in the Population Health group of CIAR were 'elders' in the Canadian health policy community, not only widely published in the academic literature, but writers of official reports, members of think-tanks, consulted and quoted by politicians and bureaucrats within Ottawa and the Provinces. When the Prime Minister created a National Forum to discuss the future of Canadian health care in 1994, four of its 24 appointees were members of the Population Health group, including the book's editor, Robert Evans. A passage in the preface to *Why are Some People Healthy and Others Not?* refers to the support:

> Intellectual as well as financial [received] from members of various governments of Canada who are concerned with health and health care policy. Both at the political and civil service level, people have

> shown a lively interest in our activities and have been very open in discussing their perceptions and concerns with us.
>
> (Evans *et al.* 1994: ix)

In sum, the members in this network are acknowledged experts of their fields, well-connected and highly influential. Although the group who wrote *Why are Some People Healthy and Others Not?* included three eminent women (Patricia Baird, Ellen Corin and Noralou Roos) the majority are men and it is tempting to blame residual patriarchy for the invisibility of women. Yet such an explanation is far too simplistic.

A singularly well-informed group, its members will have known about gender differences in mortality and morbidity rates, women's differential use of health care services, and their relatively more difficult access to the determinants of health, such as employment and other forms of economic and social capital. Their decision to ignore women cannot be explained as a matter of chance or academic absent-mindedness. At some level, conscious or unconscious, the decision was made to ignore these differences, to treat them as taken for granted, 'no longer questioned, examined or viewed as problematic' (Fujimara 1996:213).

A willingness to construct an explanation of the world which ignores women is certainly not unique to the Population Health group of CIAR or any other members in this network. Traditional ethnography, as discussed earlier in this essay, was built out of conversations with men and presented their view of their world. Seen in hindsight, one wonders how anthropologists, trained to observe and take notes on what they saw, could sit in a community yet not 'see' half its population. The equivalent mystery at the centre of this chapter is how could a group of researchers, searching all the available literature on population health, not 'see' what this literature had to say about women?

Reading the text

I took *Why are Some People Healthy and Others Not?* as my main text simply because I am more familiar with its content, but could equally have used Wilkinson's book, or the special issue of the *American Journal of Public Health.* I started by carefully checking whether my initial impression of women's relative invisibility was correct. I reviewed not only each chapter, but also the index, the tables and figures, and the bibliography. The only references in the index to women, 'gender', or 'sex', were as items under the general headings 'health status' and 'population health'. In each case, the word 'women' is bracketed with 'children'. Only two of the many figures showed data on women, both comparisons of gender differences in mortality from cardiovascular disease (Marmot and Mustard 1994). Reviewing the extensive bibliography, none of the key researchers in the epidemiology of women's health (names such as Sara Arber, Nancy Krieger, Sally Macintyre, Lois Verbrugge, Ingrid Waldron and Diana Wingard) were included.

A careful search of the text turned up an occasional sentence on male/female differences in mortality, but most passages dealing with women considered them in relation to their reproductive or mothering roles. The one exception, a chapter on social support by Ellen Corin, a medical anthropologist, made extensive use of George Brown and Tyrell Harris's (1978) *Social Origins of Depression: A Study of Psychiatric Disorder in Women.* By the conclusion of this review, my impression was that, insofar as women were represented, it was as mothers, or as depressed, or as depressed mothers, or otherwise as bodies mysteriously slow to die of heart disease. (The special issue of *Daedalus* includes an excellent chapter on Japanese women (Miyaji and Lock 1994), but as its title suggests, 'Monitoring motherhood: sociocultural and historical aspects of maternal and child health in Japan,' it is again focused on women only as mothers).

The limited use of information on women partly explains their invisibility within this text, but data are also handled in ways which obscure the presence of women. The opening chapters of the book present an impressive array of international health statistics including a series of detailed comparisons of mortality rates and income levels. These data all show that the steeper the income gradient, then the higher the mortality rate for that country. Looking for the women, I had the odd experience of sensing their presence, while being unable to see them. For it was obvious that some of the deaths, as well as some of the poverty and wealth, on which these statistics are based involved women, but how many women, or where, or when, was impossible to tell. For epidemiologists and statisticians, the aggregation of data, or their adjustment for age or sex, are simply routine procedures. This approach is so commonplace I did not question it myself until deliberately hunting for the women and finding they were missing or hidden within an aggregated data set. In their own way, these analytic techniques are as effective as is the *chador* in concealing the female presence. Everyone knows she is there, but by being veiled she becomes invisible; people act as if she has no presence or voice or relevance.

The Whitehall studies

Women are absent in a literal sense, rather than as an artifact of the statistical process, from Whitehall I, the study which sits at the core of the text *Why are Some People Healthy and Other Not?* Whitehall I is the first in a pair of studies on the health of the British civil servant. For anthropologists, there are certain key figures – such as Malinowski, Mead, Leach or Douglas – who have transformed the discipline. The epidemiologists working on the Whitehall studies are not of comparable stature, but their work is critically important to the epidemiologists, economists and others making up the Population Health network because they have provided a key piece in the puzzle linking health and inequality. To get a measure of their influence, I tried counting the citations to papers based on the Whitehall study data in the two books and the special issues of *Daedulus* and the *American Journal of Public Health*, but gave up the task as it was too time consuming. A commentary in the latter journal describes the Whitehall studies as a 'British national treasure' (Moss 1997:1413) which may seem excessive, but is an indicator of their status within this particular group of researchers.

Yet, the design of Whitehall I was fairly traditional. As described by Robert Evans (1994:5),

> (The) Whitehall study followed more than ten thousand British civil servants for nearly two decades, accumulating an extensive array of

> information on each of the individuals in the study. The data set is thus person specific and longitudinal. . . . Moreover, it is readily and unambiguously divisible into status groupings; the hierarchy of income and rank in the civil service is well-defined.

This description omits a critical point about Whitehall I; all its participants are men.

Whitehall I collected general information on mortality and morbidity rates, but its primary objective was to determine the relationship between death from CVD and rank position within the hierarchy of the civil service. In their chapter on 'Coronary heart disease from a population perspective', Marmot and Mustard summarised its main findings:

> Grade of employment shows a clear inverse relation with mortality: the lower the grade, the higher the risk. It is not only the men at the bottom who are at increased risk; each grade of employment is associated with higher risk than the one above it.
>
> (Marmot and Mustard 1994: 206)

The findings of Whitehall I describing differences in health by rank position in the British civil service have been described as 'unambiguous and large', and as 'linked to hierarchy per se, not to deprivation' (Hertzman *et al.* 1994:69).

Discussing these findings, Marmot and Mustard (1994:209) comment that, 'One is left with a fundamental puzzle as to the nature of the factor or factors that produce the gradients'. Their own answer to their own puzzle assumes that as one moves from the top of the hierarchy to the bottom, work is organised along a gradient of increasing dullness, increasing lack of control, increasing demand and increasing stress. Then drawing on stress research, particularly the work of Selye (1976), but also research by primatologists on stress and status ranking among male baboons, they argue that humans, like baboons, 'respond to a stressful environment that they cannot control with physiological changes which are dangerous to their health' (Marmot and Mustard 1994:210). The central axiom – the proposition holding the entire structure of the determinants of health model in place – is that the higher the position in the hierarchy, the lower the stress level, the better the health and the longer the life. In the remainder of this essay, I want to explore whether or not it matters that the evidence offered in support of the relationship between work, health, hierarchy and premature death which comes from the Whitehall I study excludes women.

Adding in the women

Challenging the claim that epidemiologists deal in neutral facts and objective realities, Krieger (1992:413) once wrote:

> No data bases have ever magically arrived, ready made, complete with pre-defined categories and chock full of numbers. Instead their form and content reflect decisions made by individuals and institutions, and, in the case of public health data, embody underlying beliefs and values about what it is we need to know in order to understand population patterns of health and disease.

Krieger was thinking about census data, or the sort of statistics collected by the Centre for Disease Control, but her comments apply equally well to the Population Health model. Rather than being a collection of facts, this book embodies the 'underlying beliefs and values' of those who wrote it, including the belief that gender was irrelevant to understanding population patterns of health and disease. Like the old ethnographies, their text treats women as present but invisible.

Anthropologists now know better; for example, they recognise that far more is known about the Bedouin since Abu-Lughod talked with the women and acquired a deeper understanding of how their society actually worked. The ideal method for acquiring a deeper understanding of the relationship between death and hierarchy in the British civil service might be for an anthropologist to walk the corridors of Whitehall. She could work with the women in the lower ranks, recording their observations on their job experience, listening to their commentary on the largely male world of the elite bureaucrat.

Failing an anthropologist, we have the second of the Whitehall studies, Whitehall II, in which a third of the study population are women. Whitehall II is not an ethnography of the civil service;

nevertheless, it provides some interesting insights into the position of women in the civil service hierarchy. Relatively few women work in the grades offering the greatest variety of work, the greatest degree of control and the highest levels of job satisfaction, but they make up almost two-thirds of the workers in the lower grades, where the work stress is the highest and levels of job satisfaction are the lowest. They are less likely than men to be promoted from one grade to another and promotion out of the lowest grade in the hierarchy (in which the majority of workers are women) is almost non-existent. Education and grade level are closely associated for both men and women, but with an important difference.

> Women require higher education to achieve the same grade level as men, and for a given level of education, a woman is likely to be in a lower grade than men.
>
> (Roberts *et al.* 1993)

In sum, the women of Whitehall II are more likely to be working in stressful jobs than men; they have relatively lower chances of promotion, and some will experience the additional stress of working for men less qualified than themselves.

Based on the relationship between work, stress, position in the hierarchy and mortality predicted in the model proposed by Evans (1994), one would expect that the life expectancy of the female civil servant should be significantly lower than the male. Some elements in the model hold; for example, the predicted relationships between rank position in the hierarchy, job conditions and work related stress appear to be the same in Whitehall II as Whitehall I. Men and women doing the same work, report the same degrees of job satisfaction and the same levels of psychological well-being. However, the relationships between health and position in the hierarchy are not as marked for women as they are for men, suggesting that the model of the determinants of health developed by Evans *et al.* may not hold once women are included in the study population. Whitehall II has not been in existence for long enough for good measurements of differences in mortality rates, but it is a safe assumption that women will live longer than men, although there may be differences among women. Whitehall II seems likely to create a problem for the Population Health group. If the determinants of health model are to apply to women, then it has to be capable of explaining the fundamental differences in mortality rates between men and women repeatedly documented in the epidemiological literature. The existing model must be either discarded or revised.

There are different ways of dealing with the problem of women not obeying laws based on the lives and experiences of men. Richard Wilkinson, for example, dismisses the mortality issue with a single statement, arguing that the 'causes of sex differences in survival are likely to be largely biological' (Wilkinson 1996:24). A different version of the 'women are protected by their physiology' argument proposes that women are as vulnerable to stress as men, but that their sex hormones protect their cardiovascular system until after menopause (see Chapter 4, this volume). Stress has an impact, but this is seen in other systems of the body, creating morbidity but not premature mortality. A third response is that the relationships are the same, but that women have the psychosocial advantage over men. This explanation may be phrased in terms of women being less competitive than men and therefore less stressed by their position in the hierarchy than men. Alternatively, women are protected because they are more able to access social support. Social support is credited as the surest shield against stress in *Why are Some People Healthy and Others Not?* and also as the secret of a cohesive society by Wilkinson (1996).

Evidence can be quoted for each of these explanations; all may be true, or none. Their different appeal depends partly on one's discipline and partly on one's gender perspective. My own preference inclines toward a more anthropological explanation which starts out from women's position within the structure of the civil service and ends by drawing an analogy with the Bedouin women described by Abu-Lughod. In this explanation, the women of Whitehall may be seen as living, 'in a separate community – a community that could also be considered a subsociety: separate from and parallel to the men's, yet crosscut by ties to men' (Abu-Lughod 1995:22). Their world, but also the determinants of their health, are different to those for men and cannot be explained by the same principles.

If the model developed on the basis of Whitehall I does not hold for women, is a model specific to women needed? Or, should a new model be developed which encompasses both men's and women's experiences? These questions are both theoretical and methodological. I want to look at the issues at a more practical level, by asking whether being invisible within the debate over the determinants of population health will make any difference in women's lives? Does it matter to women?

Does leaving women out affect the quality of science? The researchers who worked on Whitehall I could claim that it is their right as academics to construct whatever model of reality they choose and to analyse their data as they wish. The only criteria should be whether they could convince their peers of the inherent logic of their claims to knowledge. They might point out that Whitehall I was planned at the time when research on coronary heart disease was confined to men for a number of practical reasons, such as being able to work with smaller sample sizes and shorter periods of follow-up (see Chapter 4, this volume). As is now well acknowledged, the consequence for scientific knowledge was a narrowness of vision in which the model of the basic mechanisms of heart disease was derived from data on the male heart alone. The consequence for women was a dearth of scientific knowledge on the signs, symptoms, and best methods of diagnosing and managing heart disease in women. Physicians, trained to associate heart disease with men, tended to misdiagnose or ignore symptoms of heart disease in women. Being ignored was dangerous for women's health and bad medical science. Similarly, being invisible within the determinants of health model may prove deleterious for women and limit our understanding of the ways in which the determinants of health may function differently for women.

Setting the context

A model built exclusively on the experience of men, whether British or Bedouin, will of necessity be partial, just as traditional ethnography was partial. The dangers for women of being excluded from the clinical discourse on heart disease are now obvious, but in this final section I want to focus on the implications for women of having been excluded from the discourse on the determinants of health and on women's reactions to this exclusion.

Members in the CIAR Population Health group did not see themselves as engaged in some idle academic pursuit, but rather as producing a health policy equivalent to Keynesian economics, a guide for politicians and health policy makers. The timing of the book and the political context into which it was released contributed to their sense of its importance and impact. Health and health care were 'hot' items for the Canadian government in 1994, as evidenced by its decision to set up a National Health Forum. In its final report, that Forum commented that:

> We can say with assurance that health and health care have become a defining public issue, and Canadians have an intense interest in this debate, viewing it as a top priority for governments.
>
> (Health Canada 1997:7)

The Federal Government, while having to keep a wary eye on public opinion, had a different priority. It was committed to reducing the level of its contribution to health care. The fact that the authors, Robert Evans and his colleagues, devalued medical care relative to the impact of the social determinants of health gave the book instant appeal to many politicians and policy makers. For it could be used not only to justify cutbacks in funding for health care, but also as a rationale for transferring the money saved into job creation and deficit reduction, the twin economic mantras of the times.

Yet, the transition from a collection of academic papers into an intensely influential text was also owed to the appeal, the sheer 'transportability', of the book's language and core concepts. A working paper on population health issued by the provincial Ministers of Health in 1994 is scattered with references to the 'determinants of health' and the importance of 'a population health approach' in programme planning. Health Canada created a division labelled 'Population Health', claiming in a report that the new label denoted a significant shift in philosophical orientation away from the 'soft evidence of health promotion research' and towards the 'hard, quantitative data' used by researchers in population health (Health Canada 1996). Catching on to the new language, federal health officials quickly adopted such phrases as 'evidence based decision making', 'healthy public policy', 'health reform', and the 'determinants of health'. Responsive to government priorities, the Canadian Medical Research Council negotiated an increase in research funding, partly on the promise of setting up a committee to review research projects in the field of 'population health' (Medical Research Council of Canada 1995). The National Health Research Development Program, a government agency funding research on health, sent out calls for proposals on the determinants of health (Health Canada 1995). None of these examples are major in themselves, but the list reflects the extent to which the concepts and the language used in this book came to frame the debate over health care in Canada.

The problem for women was that this debate was now framed based on a text in which they were invisible. By ignoring them, the Population Health group seemed to have effectively placed women's issues and problems outside the new discourse on health and, therefore, outside the debate over public policy and the Canadian health care system. They only 'seemed' to have done so, however, because women ultimately refused to remain silent or invisible.

The demand to be recognised

Rather than attacking the basic framework of this discourse, women campaigned to have 'gender' included within the list of health determinants. This tactic owed a great deal to the influence of the Women's Health Bureau, a department of Health Canada charged with special responsibility for women's issues. Well-placed to lead the campaign from its position within government, the Bureau took advantage of a series of quite serendipitous events, such as its role as a joint organiser of a US/Canada conference on women's health in 1995. The Bureau inserted a workshop on 'Gender as a Determinant of Health' into the programme for that conference, encouraging Canadian debate on the issue, if also creating some bemusement among American participants, unfamiliar with the terminology, not having read the book.

Working from within Health Canada, the Bureau ensured that women's health was placed on the agenda of the National Health Forum and that a paper discussing gender as a determinant of health was included as an appendix to the Forum's final report. When the Federal Government honoured an earlier promise to set up Centres of Excellence for Research on Women's Health in 1996, the Bureau had oversight over the programme and gave priority to research on the determinants of health and their implications for women. The Bureau also played a key role in organising a conference in Halifax in 1997 on the same topic. The aim of the Bureau in undertaking each of these initiatives was to shift the parameters of the new discourse on health slightly, at least to the extent of allowing the insertion of gender within the list of health determinants.

Women's protest against their exclusion from the new discourse on health suggests that they had learned the dangers of being invisible and had resolved not to be written out of the discourse on health policy in the same way as they were once written out of the research discourse on coronary heart disease. The alliance made between women within and women outside of government was an important byproduct of this campaign and a topic which will one day require further exploration, but it is still too early to predict its evolution. Also my interest in writing this essay, was not the campaign to force the inclusion of women, but rather to understand how they had been left out of the new health discourse in the first place.

Conclusion

According to Bourdieu (1984), the power and influence of the dominant class rests in its capacity to define what is, or is not, significant. In the original formulation of the determinants of health model, gender was treated as insignificant and the model constructed out of the working experience of the male civil servant. Seen from within this model, shifting money out of medical care and other forms of social service and into economic investment made absolute sense. Information on its potential consequences for the elderly, children, the unemployed or barely employable, other vulnerable groups and women had been left out of the model and were, therefore, outside consideration.

As a final conclusion, I decided to use the story of Raven, as told by the great Haida artist, Bill Reid (1980). Raven found a clam shell on the beach, full of little creatures whom he coaxed and cajoled out into the open, the first humans. At first he found them amusing to teach and observe, but Raven became bored once 'he noticed they were all males. And no matter how hard he looked, he failed to find any females to make his games with the Haida more interesting'. Picking up a handful of molluscs, he threw them at the men, where they attached themselves briefly to their genitalia (much to the confusion of these first Haida). The men wandered off out of the story, having 'played their parts and gone their way'. Raven remained to observe the growth of the molluscs and the emergence from their shells of a new Haida people, 'brown-skinned, black-haired humans, and this time there were both males and females'. Fittingly, Bill Reid's magnificent carving of the Raven legend stands in the Museum of Anthropology at the University of British Columbia, a fitting reminder of the limitations of creating a model of the world based only on men.

References

American Journal of Public Health (1997). **87,** 1409–554.

Abu-Lughod, L. (1995). A community of secrets: the separate world of Bedouin women. In *Feminism and Community,* ed. P. Weis and M. Friedman, pp. 21–44. Philadelphia: Temple University Press.

Black, D., Morris, J., Smith, C., Townsend, P. and Whitehead, M. (1988). *Inequalities in Health: The Black Report.* London: Penguin.

Bourdieu, P. (1984). *Distinction: A Social Critique of the Judgement of Taste.* Translated by Richard Nice. Cambridge, Massachusetts: Harvard University Press.

Brown, G. W. and Harris, T. (1978). *Social Origins of Depression – a Study of*

Psychiatrc Disorder in Women. London: Tavistock Publications.

Evans, R. G. (1994). Introduction. In *Why are Some People Healthy and Others Not? The Determinants of Health of Populations*, ed. R. G. Evans, M. L. Barer and R. Marmor, pp. 3–26. New York: Aldine De Gruyter.

Franklin, S. (1997). *Embodied Progress: A Cultural Account of Assisted Conception*. London: Routledge.

Fujimura, J. (1996). *Crafting Science: A Socio-History of the Quest for the Genetics of Cancer*. Cambridge, Massachusetts: Harvard University Press.

Ginsburg, F. and Rapp, R. (1995). *Conceiving the New World Order: The Global Politics of Reproduction*. Berkeley: University of California Press.

Health Canada (1995). *Health Promotion in Canada – Case Study*. Minister of Public Works and Government Services – Canada, Catalogue #H39 – 414/1997E, Ottawa, Canada.

Health Canada (1996). *Population Health Promotion: Model of Population Health and Health Promotion*. Ottawa, Canada.

Health Canada (1997). *Final Report of the National Forum on Health*. Minister of Public Works and Government Services, Catalogue #H21–126/5–1–1997E, Ottawa, Canada.

Hertzman, C., Frank, J. and Evans, R.G. (1994). Heterogenetics in health status and the determinants of population health. In *Why are Some People Healthy and Others Not? The Determinants of Health of Populations*, ed. R. G. Evans, M. L. Barer and R. Marmor, pp. 67–92. New York: Aldine De Gruyter.

Kaufert, P. A. (1996). *Gender as a Determinant of Health*. Paper prepared for the Canada/USA Women's Health Forum, Ottawa, Ontario, August 8, 1996.

Krieger, N. (1992). The making of public health data: paradigms, politics, and policy, *Journal of Public Health Policy*, **13**, 412–27.

Lalonde, M. (1974). *A New Perspective on the Health of Canadians*, Ministry of Supply and Services. Ottawa, Canada.

Latour, B. (1987). *Science in Action: How to Follow Scientists and Engineers Through Society*. Cambridge, Massachusetts: Harvard University Press.

Latour, B. and Woolgar, S. (1986). *Laboratory Life: The Social Construction of Scientific Facts*. California: Sage.

Marmot, M. G. and Mustard, J. F. (1994). Coronary heart disease from a population perspective. In *Why are Some People Healthy and Others Not? The Determinants of Health of Populations*, ed. R. G. Evans, M. L. Barer and R. Marmor, pp.189–216. New York: Aldine De Gruyter.

McKeowan, T. (1976). *The Role of Medicine*. London: Nuffield Provincial Hospitals Trust.

McKinlay, J. B. (1993). The promotion of health through planned sociopolitical change: challenges for research and policy. *Social Science and Medicine*, **36**, 109–17.

Medical Research Council of Canada (1995). *Report of the President 1994–1995*.

Minister of Supply and Services. Ottawa, Canada.

Ministers of Health (1994). Strategies for population health: investing in the health of Canadians. Federal, Provincial and Territorial Advisory Committee on Population Health 1996. Paper prepared for the Meeting of the Ministers of Health, Halifax, Nova Scotia, September 1994. Minister of Supply and Services Canada Cat. No. #39 – 316.

Miyaji, N. T. and Lock, M. (1994). Monitoring motherhood: sociocultural and historical aspects of maternal and child health in Japan. *Daedalus*, **123**, 87–112.

Moss, N. (1997). Editorial: the body politic and the power of socioeconomic status. *American Journal of Public Health*, **87**, 1411–13.

Navarro, V. (1986). *Crisis, Health and Medicine: A Social Critique*. New York: Tavistock.

Oudshoorn, N. (1997). From population control politics to chemicals: the WHO as an intermediary organization in contraceptive development. *Social Studies in Science*, **27**, 41–72.

Proceedings of the American Academy of Arts and Sciences (1994). Health and wealth. *Daedalus*, **123**, 1–216.

Rapp, R. (1997). Constructing amniocentesis – maternal and medical discourses. In *Situated Lives: Gender and Culture in Everyday Life*, ed. H. Ragoné, P. Zavella, L. Lamphere, pp. 128–41. New York: Routledge.

Reid, R. (1980). The Haida legend of the raven and the first humans. *Museum Note No. 8*, UBC Museum of Anthropology. British Columbia, Canada.

Roberts, R., Brunner, E., White, I. and Marmot, M. G. (1993). Gender differences in occupational mobility and structure of employment in the British Civil Service. *Social Science and Medicine*, **37**, 1415–25.

Rosenberg, C. E. (1995). *Explaining Epidemics and Other Studies in the History of Medicine*, Cambridge: Cambridge University Press.

Selye, H. (1976). *The Stress of Life* (revised edition). New York: McGraw-Hill.

Stansfeld, S. A., North, F. M., White, I. and Marmot, M. G. (1995). Work characteristics and psychiatric disorder in civil servants in London. *Journal of Epidemiology and Community Health*, **49**, 48–53.

Strathern, M. (1995). *Women in Between: Female Roles in a Male World: Mount Hagen, New Guinea*. London: Rowman and Littlefield.

Wilkinson, R. G. (1996). *Unhealthy Societies: The Afflictions of Inequality*. London and New York: Routledge.

8

Agency, opposition and resistance: a systemic approach to psychological illness in sub-dominant groups

ROLAND LITTLEWOOD

The more frequent diagnosis of many mental illnesses among women, particularly married women, in most societies has been a subject for feminist critics over the last 20 years (notably Chesler 1974, Allen 1986). Is illness to be located in the individual or in her society? We might note two very general explanations:

A The pre-conditions for psychological distress in women are essentially the same as they are for men, but either because women are more vulnerable or because they are exposed to more 'stressors', they experience more illness (essentially Usher's (1991) position). But can women's experiences and political context be so easily approximated to those of men?

B The validity of the psychiatric diagnoses of women is questionable because the very notion of something like mental illness in Western societies closely relates to dominant male – and hence medical – conceptions of female identity and behaviour as characteristically 'other' (essentially Showalter's (1987) position).[1] Their illnesses then become understood as just an exaggeration of this ascription. But if so, how can women authentically act and suffer?

Very similar responses have been advocated in relation to the greater frequency of mental illness among Europe's ethnic minorities (e.g. Littlewood and Lipsedge 1982). For sub-dominant groups like women and Black people, older ideas of innate biological vulnerability to mental illness are now generally played down; yet for both groups, the idea that 'the illness' may be regarded as an intended response on the part of the patient to their situation is seldom considered – except as 'malingering' by the unsympathetic doctor. Recognition of, and resistance to, dominance is to be considered rather in some new feminist or anti-racist therapeutic response – to be as yet devised – rather than as 'the illness' itself.[2] And how can the idea that misogyny runs through both explanations be applied to the illness of men? Do they then have to be simply ignored, as Usher (1991:10) suggests?

There would certainly seem to be a problem here. Understanding any illness as a biosocial pattern requires both an interpretation of how individuals choose – or see themselves as constrained – to engage in the illness for certain ends, explicit or otherwise, together with an explanation of its pre-conditions and influences. Psychiatry, like social theory in general, remains undecided as to the relationship of the instrumental to the causal. How much do people strategically 'employ' available patterns? Alternatively, how much can one say these patterns 'reflect' particular social or biological causes? And are these two modes of thought inherently contradictory either in theory or in experience? Are moral agency and causality to be taken as additive categories, as mutually dependent, or as incommensurate? Is our very idea of agency itself cultural, perhaps masculine? In the case of those patterns conventionally termed *psychopathology*, which is more useful in examining incidence and variation: interpreting the intersubjective meanings and values which are situationally deployed by individuals out of their local cultural repertoire at a particular historical moment, whether instrumentally or otherwise; or employing such objectivised explanations as prescriptive class and gender norms, social 'stress' and lack of 'support', relative power, ecological constraints and biological imperatives? This distinction between instrumental ('pull') and causal ('push') factors is arguably one of scale as well as epistemology; it is elided, perhaps not altogether successfully, in psychoanalysis and ethology. In clinical practice, the issue is one which appears not only in medico-legal debates on responsibility but in everyday encounters when patients may see themselves as 'ill', and thus not accountable for their symptoms, while to the medical observer their actions and reported experiences may well appear motivated. And the converse.

An adequate interpretation of female illness requires an understanding of personal subjectivity and action as well as of the social field; and of male as well as female illness. In this chapter I examine, in particular, parasuicide with medically prescribed drugs, a pattern especially common among women. I suggest that we can gain an understanding of such 'overdoses' through a comparison with patterns found in less pluralistic small-scale societies; and that we look not just at the individual involved but at the local meaning of the act in the political context in which it happens. If my schema emphasises gender difference, we can replace the opposition between 'Male/Female' with one such as 'Older/Younger' (or even by 'Employed/Jobless' for in Western societies an unemployed man is, in many ways, feminised).

Overdoses: woman's violence against herself

My starting point is that contemporary women in Western societies are already identified with mood-modifying pharmaceutical tranquillisers (Jack 1992, Van der Waals *et al.* 1993). A study of such prescriptions in a North American city showed that 69% of them were given to women (Cooperstock and Sims 1971). The pattern is similar in Britain (Dunbar *et al.* 1989). Certainly, more women than men consulted doctors and received prescriptions, but there was an even greater disproportion in the number of women who were prescribed psychotropic drugs (Gabe and Williams 1986). During a single year preceding a national sampling of North American adults, 13% of the men and 29% of the women had used prescribed drugs, especially minor tranquillisers and daytime sedatives (Parry *et al.* 1973), 10% of the women in the previous two weeks (Lock 1993:397). These American rates are consistent with those of other industrialised nations. Physicians apparently expect their female patients to require a higher proportion of mood altering drugs than they do their male patients (Cooperstock 1971). These are perhaps the 'attractive healthy women who thoroughly enjoy being ill' as the *Daily Express* (1984) sardonically put it.

That we are not dealing simply with a 'real' (biological) gender disparity in psychological distress prior to social description and intervention is suggested by the symbolism of medical advertising. Stimson (1975) found that women outnumbered men by 15 to 1 in advertisements for tranquillisers and anti-depressants. One advertisement depicted a woman with a bowed head holding a dishcloth and standing beside a pile of dirty dishes represented as larger than life size; the medical consumer is then told that the drug 'restores perspective' for her by 'correcting the disturbed brain chemistry'. Employed women are rare in drug advertisements, and women are usually shown as dependent housewives and child-rearers: the world acts on them, they do not act on the world. Psychotropic drug advertisements emphasise women as passive patients. They are represented as discontented with their role in life, dissatisfied with marriage, with washing dishes or attending parent–teacher association meetings (Seidenberg 1974). The treating physician is never depicted as a woman and all the female patients appear as helpless and anxious. Advertisements for psychotropic drugs tend to portray women as patients, while those for non-psychiatric medications show men. Within the psychotropic drug category alone, women are shown with diffuse emotional symptoms, while men were pictured with discrete episodes of anxiety due to specific pressures from work or from accompanying physical illness (Prather and Fidell 1975).

Deliberate 'overdoses' of medical drugs have been said to communicate powerlessness (Hodes 1990). They are up to five times more common among the unemployed and among women, especially in the 15–19 age group (O'Brien 1986). Among girls of this age in Edinburgh during the 1970s, more than one in every hundred took an overdose each year (Kreitman and Schreiber 1979); since then, the incidence has declined somewhat (Jack 1992). Of those patients who attended hospital following a non-fatal act of deliberate self-harm, 95% had taken a drug overdose and half of the episodes involved interpersonal conflicts as the major precipitating factor (Morgan *et al.* 1975). Only a minority of women had made definite plans to prepare for death, to avoid discovery or subsequently regretted not having killed themselves. Suicidal intent and risk to life thus appear to be relatively low, especially as overdoses are usually taken with somebody close by (Hawton *et al.* 1982), for 59% occur in the presence of, or near, other people. And yet family concern is of course motivated by the fact that sometimes people who take overdoses do die, and that these individuals have generally made previous attempts to harm themselves.

Whilst the reasons given by the individual for 'taking overdoses' are often expressive (explaining the overdose as a result of personal predicament and associated feelings such as self-hatred at the time of the act), they may often be instrumental – that is they are explained in terms of the desired consequences of the act, such as the desire for increased support or understanding (Bancroft *et al.* 1979, cf. O'Brien 1986). The fact that this behaviour is a learned pattern is supported by the finding that it is concentrated in socially linked clusters of individuals (Kreitman *et al.* 1970, Chiles *et al.* 1985). A study of young women who took overdoses suggested they viewed their act as a means of gaining relief from a stressful situation or as a way of showing other people how desperate they felt. Hospital staff who assessed their motives regarded overdoses as symptomatic of psychological distress, but also noted that adolescents took overdoses in order to punish or change the behaviour of other people (Hawton *et al.* 1982). Typically a teenage girl would take tablets after a disappointment, frustration or difference of opinion with an older person (usually a parent); many patients afterwards reported that the induction of guilt in those whom they blamed for their distress was a predominant motive for the act (Bancroft *et al.* 1979). While overdoses can be seen as strategies designed to avoid or change certain specific situations, the experience of the principal is one of distress, social dislocation and extrusion; the reaction to the overdose exaggerates this extrusion, threatening exclusion from the human community altogether. Attempted suicide is, amongst other things, a dangerous adventure.

The conventional resolution of this inversion of shared life-enhancing assumptions involves its complement: medical intervention and family response endeavour to return the patient to everyday life. Not surprisingly, the overdose meets with little professional sympathy, particularly when it is perceived as having been carried out expressly as an instrumental mechanism rather than as the representation of underlying hopelessness or psychiatric illness. 'Expressive' explanations (invoking despair or depression and aiming at withdrawal, escape or death) are more acceptable and evoke more sympathy or readiness to help in both doctors and nurses than apparently pragmatic motives (Jack and Williams 1991). Doctors tend to distinguish acts as either suicidal or 'manipulative', and are more accepting of a wish to die. Nurses, themselves predominantly female, are somewhat more sympathetic than are doctors to instrumental motives; they perceive overdoses as legitimate attempts to manage distress (Ramon *et al.* 1975). Patients taking overdoses are regarded by hospital doctors as a nuisance, extraneous to the real concerns of medicine and less deserving of medical care than patients with physical illnesses, especially if the self-poisoning episode appears 'histrionic' (Hawton *et al.* 1981). As the British Medical Journal once put it: doctors 'feel a sense of irritation which they find difficult to conceal' (Anon 1971). For, to use the language of attribution theory (Shaver 1975), doctors trained in biomedicine prefer 'dispositional' to 'environmental' attributions of human suffering.

Comparative perspectives

Psychiatric research is carried out principally in urban industrialised societies where mental illness tends to be regarded by the medical profession as if it were culture free. I shall argue in this chapter, however, that certain contemporary biomedical conceptualisations and their associated patterns of social action are closely tied to the cultural politics of age and gender, particularly where these invoke notions of personal identity and attribution. Conceptualisation, therapy and the illness itself are all articulated by a shared set of values in action. Whilst the notion that biomedicine is essentially different from other theories of disease is being abandoned, there remains an assumption that Western science cannot be submitted to the type of symbolic analysis which has proved so fruitful in the study of small scale non-industrialised societies and that industrialised societies are too complex, pluralistic and pragmatic to allow us to do so. In an attempt to minimise the reduction of our subject, which includes biomedical concepts, to such concepts themselves, and to achieve a greater degree of 'universality', I shall however employ a model derived from non-Western analogues of what psychiatrists term 'psychological illness'. Where I differ from most sociological critics is in proposing a structural–functional model which allows us to consider simultaneously context, ascription, action, institution and political change.

While biomedical and 'traditional' illnesses and therapies have of course been compared before, the points of similarity have been predicated upon the primacy of the physiological or psychodynamic mechanisms of Western understandings. The converse – the direct mapping of Western categories onto traditional systems – has seldom been attempted because of the cultural contextuality of the latter as they appear to Western professionals (which has not prevented the latter from conducting the reverse procedure, that is mapping traditional categories onto Western systems, still the characteristic practice of cultural psychiatry). To attempt to identify traditional non-Western patterns in a Western population demands more than claims based on superficial phenomenological similarities that spirit possession or *amok* may occur in the West; it also requires the application to the West of the comparative models we have developed for small-scale communities. To the extent that such models are derived from within a Western academic perspective they remain culture bound but at a higher degree of universality than the clinical explanations which themselves form part of the everyday Western construction of the reactions.

'Overdoses' appear to be a relatively discrete reaction which appears, historically and geographically, culturally specific to industrialised societies, especially to the United States and Britain. While culture specific patterns may of course have a distinct biological component (as with kuru or with amphetamine psychosis), the psychiatric idea of a *culture-bound* (or *culture-specified*) *syndrome* classically described among small-scale communities has usually been taken to refer to patterns of time limited behaviour specific to a particular culture, aspects of which they exemplify and which, whilst regarded as undesirable, are recognised as discrete by local informants and medical observers alike. Few instances have a biological cause nor is the individual held to be aware or responsible in the everyday sense and the behaviour usually has a 'dramatic' quality (Littlewood and Lipsedge 1985). I shall start with a discussion of these characteristics.

Such patterns frequently articulate personal predicament but they also represent public concerns, usually what we may take as core structural oppositions between age groups or the sexes (Littlewood and Lipsedge 1985). They have a shared meaning as public and dramatic representations in an individual whose personal situation demonstrates these oppositions, and they thus generally occur in certain well-defined situations. At the same time they have a personal expressive meaning for the particular individual for whom they may be regarded as individually functional ('instrumental'): 'in situations of deprivation or frustration where recourse to personal jural power is not available, the principal is able to adjust his or her situation by recourse to "mystical pressure"' as the anthropologist Ioan Lewis (1971) puts it: that is, through appeals to values and beliefs which cannot be questioned because they are tied up with the most fundamental and taken for granted political understandings and structures of the community. How 'conscious' the principal is of pragmatically employing the mechanism as a personal strategy is debatable but it may be noted that medical observers have frequently described these reactions as 'dissociative'. Victor Turner (1969) aptly calls this side-ways recourse to tacit pressure 'the power of the weak'. The popular medical term employed when an illness cannot be seen as a biological disease is, of course, 'manipulation'.

To consider in more detail my analogous instances: *wild man behaviour* (*negi-negi*, *nenek*) is a term given to certain episodes of aggressive behaviour in the New Guinea Highlands (Newman 1964, Salisbury 1966, Koch 1968, Langness 1968, Clarke 1973, Reay 1977). The affected man rushes about erratically, threatening other people with weapons, destroying their property, blundering through the village gardens and tearing up crops. Episodes last for a few hours, or at most days, during which the wild man fails to recognise people and, on recovering, claims amnesia for the episode.

Behaviour is locally attributed to possession by spirits, and treatment may include the pouring on of water as a sort of exorcism, although Western observers have felt these measures were applied half heartedly if not theatrically (Newman 1964). The incipient wild man's initial announcement that he no longer wishes to eat and his rejection of his share of the prepared food advertises his coming performance (Clarke 1973:209). This is always public. According to Newman (1964:3), 'It would be possible for a man to run wild in seclusion but no-one does'. The audience participate by feigning terror, attempting to mollify the principal or ostentatiously hiding weapons. To Western observers he retains a high degree of control. Like the shaman in his trance, 'though he flings himself in all directions with his eyes shut [he] nevertheless finds all the objects he wants' (Eliade 1964). In *negi-negi* and similar reactions there is 'a disproportion between the injury threatened and actually inflicted. It is generally more alarming to the white onlooker than the native', the anthropologist and physician Charles Seligman wrote in 1928. (Similar episodes of spear throwing by Western Desert Aborigines remain constrained and relatively safe; only participants with organic brain disease are 'out of control' (Jones 1971).) The pattern typically occurs among young men who are politically powerless, in such hopeless situations as working to pay back a bride price debt raised through a network of their elders. *Negi-negi* exaggerates this social dislocation to the point where the young man dramatically declines membership in his social community altogether. Resolution may include latitude in repayment. The net result is to restore and legitimate the status quo, not to question it.

What we might term audience participation is essential to *negi-negi* and to certain more recognisably 'parasuicidal' behaviours. In the Western Pacific island of Tikopia, aggrieved or offended young women and men swim out to sea (Firth 1961). As an islander comments:

> A woman who is reproved or scolded desires to die, yet desires to live. Her thought is that she will go to swim, but be taken up in a canoe by men who will seek her out to find her. A woman desiring death swims to seawards; she acts to go and die. But a woman who desires life swims inside the reef.
>
> (Firth 1961:12)

While completed suicide is locally regarded as a revenge on the community, the pay-off for the survivor of a suicide swim includes enhanced status together with a re-negotiation of the original problem. Thus an adolescent girl who is rebuffed or censured by parents reacts by exaggerating this extrusion, detaching herself further from the community until the resolution ultimately restores the equilibrium. For the community, the tension between parental authority and filial independence is presented as dramatically as in the accounts of lovers' suicide pacts which regularly appear in the British popular press.

Sympathy for the Tikopian suicide swimmer wanes with repetition; like *saka* among the Kenyan Waitata, the reaction can occur 'once too often'. Approximately half of married Waitata women were once subject to *saka* (which we might gloss as 'possession') after a wish was refused by their husband, typically for an object that is a prerogative of men (Harris 1957). It was locally recognised that *saka* was clearly something to do with relations between men and women. Whilst women provided food for the family they were also expected to supply domestic objects which could only be purchased through access to the profits of the sale of land or livestock, yet such access was denied to women:

> Women are said to have no head for land or cattle transactions. They do not have the right sort of minds for important community affairs because they have little control over emotions and desires. Indeed femininity is made synonymous with an uncontrolled desire to acquire and consume. . . . In *saka* attacks, women are caricatured as uncontrollable consumers. . . . They are shown as contrasting in every way with men and the contrast is symbolised as a personal malady.
>
> (Harris 1957: 1054, 1060)

Harris suggests that 'women can acquire male prerogatives or the signs thereof through illness', and in the 1950s compared the reaction to a European women's 'sick headache' or 'pregnancy cravings'. While observers feel the reaction can be considered either 'real' or 'simulated', some local men say the whole reaction is a pretence. *Saka* is regarded by the Waitata simultaneously as an illness, as a possession by spirits, and as the consequence of a woman's personal wishes and social situation. Possession ceases when the woman's wish is granted by her husband or when he sponsors a large public ceremony in which she wears male items of dress or new clothes.

Pastoral Somali women may similarly become possessed by the *sar* spirits who demand gifts and attention (Lewis 1966, 1971); 'therapy really consists in spoiling the patient while ostensibly meeting the demands of the spirit as revealed to the expert therapist'. In a patriarchal Islamic society (before the recent civil war) in which women are excluded from the public realm, the outrageous behaviour of possession coerces Somali husbands into gestures of reconciliation and consideration while at the same time the formal public ideology of male dominance remains unchallenged. The cost of *sar* ceremonies may be such as to preclude the purchase by the husband of additional wives. The typical situation is a

> hard-pressed wife, struggling to survive and feed her children in the harsh nomadic environment, and liable to some degree of neglect, real or imagined, on the part of her husband. Subject to frequent, sudden and often prolonged absences by her husband as he follows his manly pastoral pursuits, to the jealousies and tensions of polygamy which are not ventilated in accusations of sorcery and witchcraft, and always menaced by the precariousness of marriage in a society where divorce is frequent and easily obtained by men, the Somali women's lot offers little stability or security. . . . Not surprisingly the *sar* spirits are said to hate men.
>
> (Lewis 1971:75–6)

Function and opposition

To summarise, these patterns appear to occur where major points of political and cultural oppositions are represented in a particular subdominant individual's rather drastic situation and thus, not surprisingly, where the everyday resolution or affirmation of power relationships are inadequate as perceived solutions to the problem. Such oppositions may occur between elders and younger people (*negi-negi*, Tikopian swims) or between men and women (swims, *sar*, *saka*). Employing our own psychological categories, these contentious points may represent social 'tensions'. Victor Turner (1969) says such tensions do not imply that society is about to break up; rather, they 'constitute strong unities . . . whose nature as a unit is constituted and bounded by the very forces that contend within it. [The tension becomes] a play of forces instead of a bitter battle. The effect of such a 'play' soon wears off, but the sting is removed from certain troubled relationships'. As Harris (1957:1064) comments about the *saka* attack, it 'allows a round-about acknowledgement of conflict, but in the saka dance there is again peace, dignity and festivity'; indeed key symbolic elements of this resolution are included in other rituals of the community. 'The use of symbols in ritual secures some kind of emotional compromise which satisfies the majority of the individuals who comprise a society and which supports its major institutions' (Richards 1982:169).

The patterns I have discussed may be described as demonstrating three stages. First, the individual is first extruded by others out of normal social relationships in an extension of what is already a devalued or subdominant social status.[3] This is followed by a prescribed role, deviant yet in some sense legitimate, which represents further exaggeration of this dislocation (and in this becomes a direct contravention or 'inversion' of the social values supposedly held in common by the whole community). This unbalances the social equilibrium to such an extent that it is succeeded by restitution back into conventional and now unambiguous social relationships: for a society cannot let its subdominant individuals out of the human schema altogether. 'The *saka* dance [the resolution] turns the *saka* attack on its head' (Harris 1957:1060). The Tikopia suicide voyager's 'attempt at detachment has failed, but he has succeeded in resolving his problem. He is once again absorbed and an effective catharsis has been obtained' (Firth 1961:15).

A similar three stage model of separation, transition and reintegration has been postulated for those psycho-social transitions glossed as 'rites of passage' (Van Gennep 1960), including certain types of shamanic trance possession (Peters and Price-Williams 1983) and women's cults in Africa (Wilson 1967, Turner 1969). While Western medical convention distinguishes 'symptoms' from 'treatment' within such patterns, I propose here to consider them as a single complex. As Harris (1957: 1061) notes of *saka*, '[Possession] attack and [therapeutic] dance are two manifestations of a single situation . . . [they] can be translated into one another'. Leach (1961:135–6) has suggested that what we generally recognise as ritual is 'normal social life . . . played in reverse', and he offers the terms formality, role reversal and masquerade to represent the equivalents of my three stages. The prescribed deviant role amplifies the rejection by the community, frequently taking the form of behaviour which contravenes the core values of the society, such as female modesty or decorum. During this period the principal is regarded as the victim of external mystical power (which must be placated) and is not held accountable; when in *negi-negi* 'a man does not have a name' (Langness 1965:276). He is a once domesticated creature who has now escaped from everyday social classification (Newman 1964). Whilst I am emphasising here the instrumental, rather than the expressive aspect of this behaviour, we may gloss the personal experience for the principal (and audience) by the Western psychological term 'catharsis', similar to the collective experience of social inversions found in public humour, carnivals, licensed rituals of rebellion and other contravention of norms in certain otherwise specific and tightly controlled contexts.

In large scale literate societies with, we might note, a linear rather than a cyclical expectation of time (Leach 1961), individualised reactions of 'hysterical conversion' may come to succeed the periodic and carnivalesque rites of role reversal of smaller and more homogeneous groups. Loudon (1959) describes such a 'pathologising' shift for Zulu women in the 1950s. Lee (1981) shows how analogous symbolic inversions of *amok*, *latah* and possession states continue in a now industrialised Malay society, although the attribution of responsibility has now been transferred from supernatural agencies onto an acquired Western notion of individual psychopathology.

'Functionalist' explanations of symbolic inversion such as I have offered stress both group catharsis and the social marking of a norm by its licensed and restricted contravention. The audience is placed between distress and safety, the position of 'optimal distancing' and balanced attention reserved for group catharsis (Scheff 1979). To say that the behaviour is in one understanding a 'performance' is not to suggest that the 'actors' do not fully identify with their parts, or that they are not genuinely suffering, engaged in self-discovery and perhaps exploring new identities. For the community as a whole, potential social oppositions are dramatised but also shown at the same time to be susceptible to a solution, albeit a temporary and restricted one for they are not 'ultimately' resolved. The spectacle of reversal of values in, say, *negi-negi*, articulates the structural opposition between affines and kin, wife-givers and wife-receivers, and between older and younger men. The performance embodies social oppositions cognitively and experientially for all members of the audience but also permits individual identification with the protagonist. That sex-specific performances appear to be appreciated equally by both sexes suggests that close individual identification is perhaps less important than psychodynamic writers have suggested: ritual derives its efficacy and power from its performance.

I am not however proposing an invariably homeostatic function. Diagnosis of *sar* possession is in the hands of women and its treatment groups provide an organisation for women which parallels the public practice of Islam dominated by men. The participation of women in such healing groups (which may encourage further possession but now in a solicited and controlled ritual setting) may be said to 'allow the voice of women to be heard in a male-dominated society, and occasionally enable participants to enjoy benefits to which their status would not normally entitle them' (Corin and Bibeau 1980). Ogrizek (1982) and Boddy (1989) suggest that the healing rituals associated with possession trance enable women to explore and develop an 'oppositional' female identity albeit not in evidently political terms. Such groups may partially maintain the woman's stigmatised ('sick' or 'vulnerable') identity in opposition to complete reversion to normative values, but elsewhere spirit possession therapy groups, especially those for men, may take on important roles including taxation, redistribution and welfare (Janzen 1978). Their organisation may then provide, in contrast to a hypothesised 'homeostatic' function, whether individual or communal, a dynamic mechanism for social mobility and even institutional change (Spring 1978; Janzen 1979; Lee 1981) and we can here offer parallels with 'consciousness raising' and political activism in Western women's therapy groups.

The 'mystical pressure' of Western medicine: spirits and diseases

My model of these patterns involves non-dominant individuals who display in their immediate personal situations certain basic contradictions expressed through the available intellectual tools, with recourse to 'mystical pressure' permitting personal adjustment of their situation by a limited contravention of society's core values. Each pattern involves dislocation, exaggeration to the point of inversion of the shared norms and restitution. An attempt to look for equivalent reactions in Britain would appear quite straightforward apart from the notion of 'mystical pressure'. What unquestionable 'other-worldly authority', standing outside everyday personal relations might serve to explain and legitimate them, and reduce individual accountability both in the starting situation of subdominance and in its later exaggeration? And then compel restitution?

For small-scale and non-literate communities, the categorisation of the natural world and human relations to it is referred to a fairly tight system of cosmology which we generally call 'religion'. Religion is an ideology: it both describes and prescribes, allocating the individual into the natural order. Through its other-worldly authority it legitimates personal experience and the social order. By contrast, in the secularised West, Christianity has lost its power of social regulation and competes both with other religions and, more significantly, with a variety of alternative ideologies, both moral and political. Where then can we find an equivalent 'mystical' sanction which integrates personal distress into a shared conceptualisation of the world?

I would suggest that the legitimisation of our present world view lies ultimately in contemporary science. In its everyday context as it relates to our personal experience, science is most salient in the form of medicine which offers core notions of individual agency, responsibility and action. In all societies, illness is experienced through an expressive system encoding indigenous notions of social order. Whilst serious illness 'is an event that challenges meaning in this world, . . . medical beliefs and practices organise the event into an episode which gives form and meaning' (Young 1976). The reality of an illness is derived from its ability to evoke deeply felt social responses as well as intense personal effects (Good and Good, 1981). This obligation to order abnormality is no less when it is manifest primarily through unusual behaviour. The person whose behaviour is seen as being unpredictable not only becomes an object of fear; she becomes endowed with a potentiality for a perverse form of power.

Young women who take overdoses still gain access to hospital, despite the physician's antipathy, since the popular understanding of suicidal behaviour is of 'something that happens to one', rather than something one intentionally brings about (Ginsberg 1971). The patient's family accept that her problems are outside her direct personal control, and responsibility is thereby attributed to some agency beyond the patient's volition (Ginsberg 1971, cf. James and Hawton 1985). Thus we have the continuing popular use of the (quasi-accidental) 'overdose' as opposed to active 'self-poisoning' or 'attempted suicide'.

Professional intervention in a sickness involves incorporating the patient into an overarching system of explanation, a common structural pattern which manifests itself in the bodily economy of every human being. Accountability is transferred onto an agency beyond the patient's control, diseases rather than spirits. Both disease and spirits are 'other', non-social. Both compel the intervention of other people. Becoming sick is part of a social process leading to public recognition of an abnormal state and a consequent readjustment of patterns of behaviour and expectations, and then to changed roles and altered responsibility. Expectations of the sick person include exemption from discharging some social obligations, exemption from responsibility for the condition itself, together with a shared recognition that it is undesirable and involves an obligation to seek help and co-operate with treatment (Parsons 1951). Withdrawal from everyday social responsibilities is rendered acceptable through some means of exculpation, usually through mechanisms of biophysical determinism:

> When faced with a diagnosis for which he has equally convincing reasons to believe that either his client is sick or he is not sick, the

> physician finds that the professional and legal risks are less if he accepts the hypothesis of sickness.
>
> (Young 1976)

To question the biomedical scheme itself involves questioning some of our most fundamental assumptions about nature and human agency. Because of its linking of personal experience with the social order, its standardised expectations of removing personal responsibility and initiating an institutionalised response, and its rooting in ultimate social values through science, biomedicine offers a powerful and unquestionable legitimate inversion of everyday behaviour – as a disease. It will thus not be surprising to find many non-Western equivalents already included by medicine in psychiatric nosologies. Others, like chronic pain, chronic fatigue, 'factitious' illness and eating disorders, we may suspect lie hidden more effectively in the fringes of physical medicine rather than in psychiatries.

Why women?

My model emphasises a discrepancy in power and opportunity between the dominant and the subdominant groups within a community. In the vast majority of societies women are 'excluded from participation in or contact with some realm in which the highest powers of the society are felt to reside' (Ortner 1974). They are excluded by a dominant ideology which reflects men's experiences and immediate interests. 'The facts of female physiology are transformed in almost all societies into a cultural rationale which assigns women to nature and the domestic sphere, and thus ensures their general inferiority to men' (La Fontaine 1981:347). The core aspects of the female role in Western society are reflected in the ideals still held out to women: marriage, home and children as the primary focus of concern, with reliance on a male provider for sustenance and status. There is an expectation that women will emphasise nurturance and that they live through and for others rather than for themselves. Women are still often expected to give up their occupation and place of residence when they marry and are banned from the direct assertion and expression of aggression. The bodies of women of reproductive age are biomedically 'normal', against which their later menopause is a pathology (Lock 1993), yet through childbearing every woman in the West becomes a potential patient. Their lack of power is attributed to their greater emotionality and their inability to cope with wider social responsibilities, for dependency and passivity are expected of a woman; her image in clinical psychology tests is of a person with a childish incapacity to govern herself and a need for male protection and direction (Broverman *et al.* 1970). Contemporary Western women are permitted greater freedom than men to 'express feelings' and to recognise emotional difficulties (Phillips and Segal 1969), enabling the women to define her problems within a medical framework and bring them to the attention of her doctor (Gove and Tudor 1973, Horwitz 1977).

Jordanova (1980) has suggested that medical science is characterised by the action of men on women; women are regarded as more 'natural', passive, awaiting male ('cultural') organisation. As a British gynaecology textbook put it not so long ago, 'Femininity tends to be passive and receptive, masculinity to be more active, restless, anxious for repeated demonstrations of potency' (James 1963). Clearly not all doctors are men (indeed, medicine has a greater proportion of women than other professions) but the woman doctor's role in relation to her patient replicates the male–female relationship: a male patient may flirt with his nurse but hardly with his doctor. To decide when 'complementarity' becomes 'dominance', 'opposition', 'tension' or 'conflict' depends on our understanding of individual action lived through such a schema.[4]

The normative situation of active male (husband, older doctor) and passive female (wife, younger patient) is reflected in the drama of the hospital casualty department. The unease and anger which the overdose evokes in the medical profession reflect its 'perverse' transformation of the clinical paradigm. The official translation of the behaviour into symptoms takes place under socially prescribed conditions by the physician who alone has the power to legitimate exculpating circumstances. As with nineteenth century hysteria or *saka*, the resolution of the reaction invokes a 'mystical pressure' which simply replicates the social structure in which the reaction occurs[5]; like other 'culture-bound syndromes' it displays core structural antagonisms but shows they are 'soluble' within the existing political and symbolic framework. As with possession cults, overdoses can sometimes become collective with the possible development of an oppositional or strategic perspective (e.g. *Times* 1990; compare suffragettes or other political prisoners on hunger strike refusing food). The drama of the overdose scene in the casualty department replays the male doctor–female patient theme without questioning it, but it does afford a degree of negotiation for the principal, who induces a mixture of responses, but particularly guilt, in family and friends (James and Hawton 1985); to paraphrase Engels, women act but not under conditions of their own choosing.

Ideas of 'structure' and 'function' are currently unfashionable in social anthropology because we now recognise these as external models independent of peoples' own interpretations of their experience and possibilities. Structure is not an empirical, and seldom a subjective, reality (and perhaps a more appropriate term for my model would be a 'map'[6]). Certainly, a cybernetic schema of the sort I have outlined leaves individual motivation a little mysterious (Geyer and Van der Zouwen 1986). Actually how 'conscious' (and thus for us, how 'responsible') is the principal? I would suggest that we can never answer such a question empirically, for the local perception of agency – or lack of agency – is itself an essential element in the model itself. Ambiguity over conscious interest is built into the whole pattern, and may be found even in the spirit possession instances, as well as in those patterns, like Tikopian suicide trips and overdoses, which employ more internalised mystical pressure such as 'psychological illness' or 'pressure' or 'female personality', and which thus come somewhat closer to a recognition of the individual's agency (and only through which may our patterns then become locally explicit as political resistance). Nor am I arguing that my model represents a neat homeostatic system in which each extrusion is automatically followed by exaggeration and restitution. The principal may – perhaps generally does – remain in the subdominant or extruded position, in a hopeless surrender to authority or in a perhaps more comfortable internalisation of her position.[7] If my map suggests core cultural themes and practical possibilities implicitly and often explicitly 'recognised' by a culture, its workings out may invoke tentative attempts and aesthetic play within the theme, in microsituations and misinterpretations, toyed with, fantasised, presumed, hinted at, as well as in overt distress and disaster; for, unless the restitution occasionally fails, mystical pressure ceases to maintain its power. The overdose is socially compelling because, every now and then, we learn that a woman dies.

Endnotes

1 To be fair, both Showalter and Usher consider the two positions but the former examines more closely the relationship between diagnosis and historical context. Position A (vulnerability and differential stress) is also that of most medical commentators, B of most feminist sociologists who invoke 'social construction' theories.

2 Nineteenth century hysteria is perhaps an exception: see Showalter's (1993) critique of recent Lacanian proposals that the Victorian hysteric demonstrated a bodily proto-resistance – aware or more usually implicit – to patriarchal 'logocentrism'.

3 Ortner, 1974; although many 'second wave' identity feminists – Cixous, Chodorow, Daly, Dworkin, Irigary, Rich (cf. Kristeva) have re-essentialised all women as biologically 'closer to nature': nurturant and emotionally authentic, not driven by the male quest for power over one's fellows.

4 Whether 'the charisma of dominance comes from a particular power – that of ultimately defining the world in which non-dominants live – to reveal it will require more than the examination of crude, arbitrary cruelties or exploitations' (Ardener 1989:186).

5 Indeed Bryan Turner (1984) has proposed the term 'sacred disease' for those Western patterns of psychological illness which are represented by male control over women and which thus represent the essence of social power within the community.

6 As Bateson put it, comparing the map to the 'territory' (reality).

7 Women who take overdoses which do not resolve their situation may be left even more powerless and now with 'internally directed hostility' (Sakinofsky *et al.* 1990). The same phenomenon is susceptible of a variety of interpretations: reflection of social misogyny; hopeless identification with overwhelming 'stress'; attempt to assert control over trauma by mimesis; social catharsis or homeostatis; cultural loophole; individual catharsis; role reversal; theatre; entertainment; ritual reaffirmation of gender relationships; rite of passage; genesis of sorority; organised resistance; revolutionary prototype; expression or resolution of symbolic ambiguity; calculated motivation; not to mention the enactment of such individual cognitions as distress, parody, play, adventure or revenge. And

even, should we wish, biological vulnerability. I suggest that overdoses may be understood in any of these ways, or indeed all of them. Claims to the primacy of a particular interpretation ultimately remain arbitrary, grounded not only in the immediate outcome but in the observer's own academic, professional and political assumptions.

Acknowledgements

My model here was first put forward with Maurice Lipsedge (Littlewood and Lipsedge 1987) and elaborated in the case of overdoses in Symptoms, Struggles, Functions (Littlewood 1994).

References

Allen, H. (1986). Psychiatry and the construction of the feminine. In *The Power of Psychiatry*, ed. P. Miller and N. Rose, pp. 85–110. Cambridge: Polity Press.

Anononymous. (1971). Suicide attempts: editorial. *British Medical Journal*, **11**, 483.

Ardener, E. (1989). The problem of dominance. In *The Voice of Prophecy*, ed. E. Ardener, pp. 127–43. Oxford: Blackwell.

Bancroft, J., Hawton, K., Simpkins, S., Kingston, B., Cumming, C. and Whitwell, D. (1979). The reasons people give for taking overdoses. *British Journal of Medical Psychology*, **52**, 353–65.

Boddy, J. (1989). *Wombs and Alien Spirits: Women, Men and the Zâr Cult in Northern Sudan*. Madison: University of Wisconsin Press.

Broverman, I. D., Broverman, D. M., Clarkson, F. E., Rosenkrantz, P. S. and Vogel, S. R. (1970). Sex role stereotypes and clinical judgements of mental health. *Journal of Consulting and Clinical Psychology*, **34**, 1–7.

Chesler, P. (1974). *Women and Madness*. London: Allen Lane.

Chiles, J. A., Strosahl, K. D., McMurtray, L. and Lineham, M. M. (1985). Modelling effects on suicidal behaviour. *Journal of Nervous and Mental Disease*, **173**, 479–81.

Clarke, W. C. (1973). Temporary madness as theatre: wild-man behaviour in New Guinea. *Oceania*, **43**, 198–214.

Cooperstock, R. (1971). Sex differences in the use of mood-modifying drugs. *Journal of Health and Social Behaviour*, **12**, 238–44.

Cooperstock, R. and Sims, M. (1971). Mood-modifying drugs prescribed in a Canadian city. *American Journal of Public Health*, **61**, 1007–16.

Corin, E. and Bibeau, G. (1980). Psychiatric Perspectives in Africa: 2. *Transcultural Psychiatric Research Review*, **16**, 147–78.

Daily Express. (1984). Wives hooked on illness are giving GPs a headache. 27th March, London.

Dunbar, G. C., Perera, M. H. and Jenner, F. A. (1989). Patterns of benzodiazepine use in Great Britain. *British Journal of Psychiatry*, **155**, 836–41.

Eliade, M. (1964). *Shamanism: Archaic Techniques of Ecstasy*. Princeton: Princeton University Press.

Firth, R. (1961). Suicide and risk-taking in Tikopian society. *Psychiatry*, **2**, 1–17.

Gabe, J. and Williams, P. (1986). *Tranquillisers: Social, Psychological and Clinical Perspectives*. London: Tavistock.

Geyer, F. and Van der Zouwen, J. (1986). *Sociocybernetic Paradoxes: Observation, Control and Evolution of Self-Steering Systems*. London: Sage.

Ginsberg, G. P. (1971). Public conceptions and attitudes about suicide. *Journal of Health and Social Behaviour*, **12**, 200–1.

Good, B. J. and Good, M. -J. D. (1981). The meaning of symptoms: a cultural hermeneutic model for clinical practitioners. In *The Relevance of Social Science for Medicine*, ed. L. Eisenberg and A. Kleinman, pp. 165–201. Dordrecht: Reidel.

Gove, W. R. and Tudor, J. F. (1973). Adult sex roles and mental illness. *American Journal of Sociology*, **78**, 812–35.

Hage, P. and Harary, F. (1983). *Structural Models in Anthropology*. Cambridge: Cambridge University Press.

Harris, G. (1957). Possession 'hysteria' in a Kenyan tribe. *American Anthropologist*, **59**, 1046–66.

Hawton, K., Marsack, P. and Fagg, J. (1981). The attitudes of psychiatrists to deliberate self-poisoning: comparison with physicians and nurses. *British Journal of Medical Psychology*, **54**, 341–8.

Hawton, K., Osborne, M. and Cole, D. (1982). Adolescents who take overdoses. *British Journal of Psychiatry*, **140**, 118–23.

Hodes, M. (1990). Overdosing as communication: a cultural perspective. *British Journal of Medical Psychology*, **63**, 319–33.

Horwitz, A. (1977). The pathways into psychiatric treatment: some differences between men and women. *Journal of Health and Social Behaviour*, **18**, 169–78.

Jack, R. (1992). *Women and Attempted Suicide*. Hove: Erlbaum.

Jack, R. L. and Williams, J. M. G. (1991). Attribution and intervention in self-poisoning. *British Journal of Medical Psychology*, **64**, 359–73.

James, C. W. B. (1963). Psychology and gynaecology. In *British Gynaecological Practice*, ed. A. Cloge and A. Bourne. London: Heinemann.

James, D. and Hawton, K. (1985). Overdoses: explanations and attitudes in self-poisoners and significant others. *British Journal of Psychiatry*, **146**, 481–5.

Janzen, J. M. (1979). Drums anonymous: towards an understanding of structures of therapeutic maintenance. In *The Use and Abuse of Medicine*, ed. D. M. Vries, pp. 154–67. New York: Praeger Scientific.

Jones, I. H. (1971). Stereotyped aggression in a group of Australian Western Desert Aborigines. *British Journal of Medical Psychology*, **44**, 259–65.

Jordanovna, L. (1980). *Sexual Visions: Images of Gender in Science and Medicine Between the 18th and 19th Centuries*. Hemel Hempstead: Harvester.

Koch, F. (1968). On 'possession' behaviour in New Guinea. *Journal of the Polynesian Society*, **77**, 135–46.

Kreitman, N. and Schreiber, M. (1979). Parasuicide in young Edinburgh women. *Psychological Medicine*, **9**, 469–79.

Kreitman, N., Smith, P. and Tan, E. (1970). Attempted suicide as language. *British Journal of Psychiatry*, **116**, 465–73.

La Fontaine, J. (1981). The domestication of the savage male. *Man* (n.s.), **16**, 333–49.

Langness, L. L. (1968). Hysterical psychosis in the New Guinea Highlands: a Bena Bena example. *Psychiatry*, **28**, 258–77.

Leach, E. R. (1961). *Rethinking Anthropology*. London: Athlone.

Lee, R. L. M. (1981). Structure and anti-structure in the culture-bound syndromes: the Malay case. *Culture, Medicine and Psychiatry*, **5**, 233–48.

Lewis, I. M. (1966). Spirit possession and deprivation cults. *Man* (n.s.), **1**, 307–29.

Lewis, I. M. (1971). *Ecstatic Religion*. Harmondsworth: Penguin.

Littlewood, R. (1994). Symptoms, struggles, functions. In *Gender, Drink and Drugs*, ed. M. McDonald, pp. 77–98. Oxford: Berg.

Littlewood, R. and Lipsedge, M. (1982). *Aliens and Alienists: Ethnic Minorities and Psychiatry*. Harmondsworth: Penguin.

Littlewood, R. and Lipsedge, M. (1985). Culture-bound syndromes. In *Recent Advances in Psychiatry – 5*, ed. K. Granville-Grossman, pp. 105–26. Edinburgh: Churchill Livingstone.

Littlewood, R. and Lipsedge, M. (1987). The butterfly and the serpent. *Culture, Medicine and Psychiatry*, **11**, 289–335.

Lock, M. (1993). *Encounters with Aging: Mythologies of the Menopause in Japan and North America*. Berkeley: University of California Press.

Loudon, J. R. (1959). Psychogenic disorders and social conflict among the Zulu. In *Culture and Mental Health*, ed. M. K. Opler, pp. 351–69. New York: Macmillan.

Morgan, H. G., Burns-Cox, C. J., Pocock, H. and Pottle, S. (1975). Deliberate self-harm: Clinical and socio-demographic characteristics of 368 patients. *British Journal of Psychiatry*, **127**, 574–9.

Newman, P. L. (1964). 'Wild man' behaviour in a New Guinea Highlands community. *American Anthropologist*, **66**, 1–19.

O'Brien, S. (1986). *The Negative Scream: A Story of Young People Who Took an Overdose*. London: Routledge and Kegan Paul.

Ogrizek, M. (1982). Mama wata: de l'hystérie à la feminité en Afrique. *Confrontations Psychiatriques*, **21**, 213–37.

Ortner, S. B. (1974). Is female to male as nature is to culture? In *Women,*

Culture and Society, ed. M. Rosaldo and L. Lamphere, pp. 67–88. Stanford: Stanford University Press.

Parry, H. J., Balter, M. D., Mellinger, G. D., Cisin, I. H. and Manheimer, D. I. (1973). National patterns of psychotherapeutic drug use. *Archives of General Psychiatry*, **28**, 769–83.

Parsons, T. (1951). Illness and the role of the physician: A sociological perspective. *American Journal of Orthopsychiatry*, **21**, 452–60.

Peters, L. G. and Price-Williams, D. (1983). A phenomenological overview of trance. *Transcultural Psychiatric Research Review*, **20**, 5–39.

Phillips, D. L. and Segal, B. (1969). Sexual status and psychiatric symptoms. *American Sociological Review*, **29**, 678–87.

Prather, J. and Fidell, L. (1975). Sex differences in the content and style of medical advertisements. *Social Science and Medicine*, **9**, 23–6.

Ramon, S., Bancroft, J. H. J. and Skrimshire, A. M. (1975). Attitudes towards self-poisoning among physicians and nurses in a general hospital. *British Journal of Psychiatry*, **127**, 257–64.

Reay, M. (1977). Ritual madness observed: a discarded pattern of faith in Papua New Guinea. *Journal of Pacific History*, **12**, 55–79.

Richards, A. (1982). *Chisungu: A Girl's Initiation Ceremony Among the Bemba of Zambia*. London: Tavistock.

Sakinofsky, I., Roberts, R. S., Brown, Y., Cumming, C. and James, P. (1990). Problem resolution and repetition of parasuicide. *British Journal of Psychiatry*, **156**, 395–9.

Salisbury, R. (1966). Possession in the New Guinea Highlands. *Transcultural Psychiatric Research Review*, **3**, 103–8.

Scheff, T. J. (1979). *Catharsis in Healing, Ritual and Drama*. Berkeley: University of California Press.

Seidenberg, R. (1974). Images of health, illness and women in drug advertising. *Journal of Drug Issues*, **4**, 264–7.

Seligman, C. G. (1928). Anthropological perspectives and psychological theory. *Journal of the Royal Anthropological Institute*, **62**, 193–228.

Shaver, K. G. (1975). *An Introduction to Attribution Processes*. Cambridge, Massachussetts: Winthrop.

Showalter, E. (1987). *The Female Malady: Women, Madness and English Culture 1830–1980*. London: Virago.

Showalter, E. (1993). Hysteria, feminism and gender. In *Hysteria Beyond Freud*, ed. S. L. Gilman, H. King, R. Porter, G. S. Rousseau and E. Showalter, pp. 286–304. Berkeley: University of California Press.

Spring, A. (1978). Epidemiology of spirit possession among the Luvale of Zambia. In *Women in Ritual and Symbolic Roles*, ed. J. Hoch-Smith and A. Spring, pp. 165–78. New York: Plenum.

Stimson, G. (1975). Women in a doctored world. *New Society*, **32**, 265–6.

The Times (1990). (untitled, on collective overdoses) 12 June. London.

Turner, B. (1984). *The Body and Society: Explorations in Social Theory*. Oxford: Blackwell.

Turner, V. (1969). *The Ritual Process*. London: Routledge and Kegan Paul.

Usher, J. (1991). *Women's Madness: Misogyny or Mental Illness?* Hemel Hempstead: Harvester Wheatsheaf.

Van der Waals, F., Mohrs, J. and Foets, M. (1993). Sex differences among recipients of benzodiazepines in Dutch general practice. *British Medical Journal*, **307**, 363–6.

Van Gennep, A. (1960). *The Rites of Passage*. London: Routledge and Kegan Paul.

Wilson, P. J. (1967). Status ambiguity and spirit possession. *Man* (n. s.), **2**, 366–78.

Young, A. (1976). Some implications of mental beliefs and practices for social anthropology. *American Anthropologist*, **78**, 5–24.

Glossary

adrenal glands endocrine glands situated above the kidneys, composed of two distinct parts, the adrenal medulla, which secretes adrenaline and noradrenaline (also known as epinephrine and norepinephrine), and the adrenal cortex, which secretes corticosteroids, including cortisol.

acephalic societies do not possess a centralised political authority, usually characteristic of hunting and gathering or pastoralist societies.

androgens 'male' sex hormones.

atherosclerosis a change in the lining of arteries consisting of an accumulation of lipids, complex carbohydrates, blood and blood products, fibrous tissue and calcium deposits. Can lead to a heart attack or stroke when narrowed arteries supplying the heart or brain become completely blocked by a blood clot.

cardiac output the amount of blood pumped out of the heart each minute.

diastolic blood pressure the lower blood pressure reading, being the pressure of blood in the arteries between heart beats.

emic perspective an ethnographic representation that uses concepts and categories for analysis derived from the culture under study (compare to etic perspective).

endocrine glands hormone secreting glands.

endogamy a rule requiring marriage within a specified social or kinship group.

endogenous hormones hormones secreted in the body.

epidemiology the study of the distribution and causes of disease in human populations.

etic perspective an ethnographic representation that uses concepts and categories for analysis derived from the anthropologist's culture (compare to emic perspective).

exogamy a rule requiring marriage outside one's own social or kinship group.

exogenous hormones hormones made outside the body, such as those taken in hormone replacement therapy.

female circumcision also referred to as female genital mutilation (FGM), ritual cutting of the female genitalia, practiced mostly in African societies, that may result in complications such as reproductive tract infections and infertility. It is a deeply rooted cultural practice but there is increasingly world-wide concern about its negative consequences.

foragers people surviving by a mode of subsistence based on hunting, fishing, collecting and gathering of wild animals and plants.

height-for-age measure of an individual's growth status (height in metres relative to age), which largely reflects the adequacy of past, as opposed to present, environmental conditions.

hunter–gatherers see foragers.

hypertension high blood pressure, generally designated according to fairly arbitrary cut-off points for systolic and diastolic blood pressure.

hypothalamus the region of the brain which exerts overall control over the sympathetic nervous system.

inclusive fitness an individual's own reproductive success plus his or her effects on the reproductive success of his or her relatives, each one weighted by the appropriate coefficient of relatedness (e.g. half for a sibling, because half of siblings' genes are typically shared).

infant the earliest stage of childhood, applied by demographers to a child under one year of age .

lactational amenorrheoa suppression of the menstrual cycle during lactation or breast feeding.

life expectancy at birth indicates the number of years a newborn infant would live if patterns of mortality prevailing at the time of its birth were to stay the same throughout its life.

low-density lipoprotein (LDL) lipids are combined with proteins in the blood, to form lipoproteins, which vary in the amounts of cholesterol they contain; cholesterol is carried mainly in low-density lipoprotein and a high level of LDL indicates a greater risk of cardiovascular disease because of its association with atherosclerosis.

lymphatic filariasis infection by a tropical worm transmitted by a mosquito – it grows in the lymphatic vessels and after years of repeated infection the legs or the genitals may become grossly thickened (elephantiasis).

matrilineal descent a form of descent whereby people trace their primary kin connections through their mothers.

medicalisation the translation of social phenomena (such as poverty) or biological events (such as menstruation) into medical terminology and concepts.

menarche the first menstrual bleeding, or period.

meta-analysis an analysis of results from a number of comparable studies.

morbidity illness.

mortality death.

Murdock and White's 'standard cross cultural sample' a sample of different cultures designed to provide a good representation of the world's culture types.

myocardium the heart muscle.

neonate a new-born baby, conventionally limited to the first 4 weeks of life.

onchocerciasis (also known as river blindness) a worm infestation and form of filariasis which is transmitted by flies which live near fast-running streams.

onchodermatitis thickening of the skin associated with onchocerciasis.

pathological showing evidence of disease.

patrilineal descent a form of descent whereby people trace their primary kin relationships through their fathers.

phenomenology a movement within the social sciences which is concerned with the study of consciousness and the way in which people understand the world around them, a central notion is that objects do not exist apart from in the manner in which they are constituted by consciousness.

platelets fragments of blood cells, primarily involved in clotting.

pluralistic society a form of society in which members of sub-groups maintain independent traditions.

pruritis itching.

psychosurgery any operation on the brain carried out as a treatment for mental illness.

psychotropic drugs drugs that have an effect on the mind, including sedatives, sleeping pills and tranquillisers.

reproductive success the number of offspring an individual produces which reach reproductive age; an individual's genetic contribution to the next generation, as compared to the contributions of other individuals.

schistosomiasis (also known as bilharzia) caused by infection by a small fluke which spends part of its life-cycle in water-snails and part in humans.

sex ratio the number of males relative to the number of females in a population, usually (males/females × 100) so that a sex ratio over 100 indicates more males than females and under 100 indicates more females than males.

stunting short height in response to poor environmental conditions such as undernutrition or infection, measured as low height-for-age.

subsistence societies societies which produce enough food and other material resources for their own survival and perpetuation without the necessity for monetary exchange.

sympathetic nervous system one of the two divisions of the autonomic nervous system (the other is the parasympathetic nervous system), which are responsible for the automatic, unconscious regulation of bodily function, it controls the secretion of adrenaline by the adrenal medulla and cardiac output.

systolic blood pressure the higher blood pressure reading, being the pressure in the arteries when the heart beats.

total fertility rate total number of live births per woman.

total peripheral resistance resistance to blood flow from blood vessels – increased resistance leads to higher blood pressure.

vectors animals, often insects, which transmit disease organisms from person to person or from other infected animals to humans, e.g. mosquitoes are vectors of malaria and yellow fever.

virilocal residence a residence pattern in which after marriage a couple lives with or near the man's family or kin group.

wasting low weight resulting from breakdown of fat and/or muscle, due to starvation.

weight-for-age measure of an individual's growth status (weight in kilograms relative to age).

weight-for-height measure of an individual's growth status (weight in kilograms divided by height in metres) which is independent of age.

X-linked recessive disorders diseases caused by genes on the X-chromosome – because males only receive one copy of most genes on the X-chromosome they are much more likely to suffer from diseases caused by recessive genes than are females.

Index